ELECTRONIC BASIS OF THE STRENGTH OF MATERIALS

This book is the first to relate the complete set of strength characteristics to the electronic structures of the constituent atoms. These relationships require knowledge of both the chemistry and physics of materials. Also, the book uses both classical and quantum mechanics since both are needed to describe the properties of atoms. The book begins with short reviews of the two mechanics. Following these reviews, the three major branches of the strength of materials are given their own sections. They are: the elastic stiffnesses; the plastic responses; and the nature of fracture.

Elastic deformation can be reduced to two pure types: volume changes, and shape changes (shears). The moduli (stiffnesses) associated with each of these can be quantitatively obtained by means of the Heisenberg Principle and the theory of polarizability. The analytic theories are simple enough to indicate the physical origins of these properties. The most important atomic properties are the valence electron densities, and the electronic polarizabilities. These lead to electron exchange forces, and photon exchange forces, respectively. Atomic, molecular, and plasmonic polarizabilities play important roles. The anomalously large shear stiffness of diamond is explained in terms of the distribution of electrons along the covalent bonds, resulting from electron correlation.

For plastic deformation, the critical role of dislocation mobility is discussed. In nearly perfect metal crystals (and ionic salts) there is no quasi-static resistance to dislocation motion (only electron and phonon viscosity). Resistance in imperfect crystals is caused by extrinsic factors. In nearly perfect covalent crystals (semiconductors), dislocation mobility is intrinsically limited by the localized chemical bonds. Using diamond as the prototype, a quantitative theory of intrinsic mobility is presented.

Fracture is discussed in terms of intrinsic surface energies and their effects on the fracture surface energies of real materials. Two simple approaches are discussed. One is based on the Heisenberg Principle, the other on the plasmon theory of Schmit and Lucas. The latter may be important for understanding ductility.

This book presents a comprehensive view of the relationships between the electronic structures of solids and the microscopic and macroscopic mechanical properties of solid materials. It will be of great value to academic and industrial research workers in the sciences of metallurgy, ceramics, micro-electronics, and polymers. It will serve well as a supplementary text for the teaching of solid mechanics.

ELECTRONIC BASIS OF THE STRENGTH OF MATERIALS

JOHN J. GILMAN
UCLA, Los Angeles, CA 90095

CAMBRIDGE
UNIVERSITY PRESS

CAMBRIDGE UNIVERSITY PRESS
Cambridge, New York, Melbourne, Madrid, Cape Town, Singapore, São Paulo

Cambridge University Press
The Edinburgh Building, Cambridge CB2 8RU, UK

Published in the United States of America by Cambridge University Press, New York

www.cambridge.org
Information on this title: www.cambridge.org/9780521620055

First published 2003
This digitally printed version 2008

A catalogue record for this publication is available from the British Library

Library of Congress Cataloguing in Publication data

Gilman, John J. (John Joseph)
Electronic basis of the strength of materials / John J. Gilman.
p.   cm.
Includes bibliographical references and index.
ISBN 0 521 62005 8
1. Strength of materials.   2. Electronic structure.   I. Title.
TA405 .G54   2003
620.1′12 − dc21   2002073591

ISBN 978-0-521-62005-5 hardback
ISBN 978-0-521-07894-8 paperback

Contents

Preface

In the middle of the twentieth century it was recognized that the theory of the mechanics of continua (particularly solid mechanics) is not adequate to account for the strength properties of materials. It was clear that structural considerations at the microscopic and nanoscopic levels of aggregation are not only important but are essential to an understanding of the strengths of materials. It also came to be realized, as a result of the fact that plastic shear deformation is usually heterogeneous, that a space can be continuous but not simply connected, that is, dislocated. Also, in crystals, such dislocations are quantized, having constant displacements equal to the magnitudes of translation vectors of the crystal structure.

Although structural geometry plays an essential role in determining mechanical behavior, it leaves a number of questions unanswered. The answers to these questions can only be found by considering the electronic structure that underlies the geometric factors. Since the behavior of electrons is not described by classical mechanics, this necessitates the use of quantum mechanics to obtain answers to the various unanswered questions. For example, why is the shear stiffness of diamond greater than its volumetric stiffness? Why do the most simple metals, the alkalis, have body-centered cubic crystal structures which are not atomically close packed? Why is pure silicon brittle, while pure nickel is quite ductile? They both have the same crystal structure, so why is pure TiC hard, while pure NaCl is soft? Why do intermetallic compounds often change their color when they are plastically deformed? What determines the temperature dependences of yield stresses? Why are grain boundaries weak? What activates plastic flow and crack growth at low temperatures where there are no thermal fluctuations? Only the subatomic behavior of electrons can account for these phenomena. This is the theme of this book.

I recall seeing a review of a book (*Electronic Structure and the Properties of Solids*) by Walter Harrison that called it "idiosyncratic". It is that, but it is also comprehensive and instructive as only a lifetime of scholarship could make it. This book is also idiosyncratic. That is, it is one person's view of a vast and complex subject. Whether it is instructive is a verdict to be brought by readers.

This book differs from most books dealing with electronic structure which introduce the formalities of quantum mechanics, and then circumvent them with approximations and numerical computations. Here, it is assumed that solids are so complex that it is better to make the approximations first, and then try to show that they are consistent with the basic rules of quantum mechanics, and that they yield properties consistent with the measured values. In this regard much use is made of Heisenberg's Theorem in its exact form, so there is nothing uncertain about it. It is an approximate solution of Schrödinger's equation.

Perhaps the most controversial topic in the book is that of dislocation mobility. Together with W.G. Johnston, the author made the first direct measurements of dislocation mobility, and therefore has some knowledge of its intricacies. The latter have not been described consistently in much of the extant literature. A purpose of this book is to correct this.

Sometimes explicitly, and often implicitly, the electronic properties on which the strength properties depend are the valence charge density, and the electronic polarizability. The reader should have a better appreciation of this after reading this book. Since these are also the electronic properties that primarily determine the optical and the chemical properties, this demonstrates the unity that exists among the various properties.

Section I

Introduction

The central property of a solid is its strength. Upon this property several engineering disciplines have been built. First came the construction of shelter, and of ordnance for hunting and war, as well as tools for domestic use. Then came civil, mechanical, electrical, and chemical engineering; followed by electronics, aerospace, and now micro-device, optical, and biomedical engineering. Major civilizing events have resulted from advances in the strength of materials. The most dramatic was the industrial revolution which resulted directly from the jump in the strength of structures introduced by inexpensive steel. Before that, mining, weaving, military ordnance, agriculture, sailing, and various crafts advanced as material strengths increased. Inexpensive steel transformed the ways in which bridges, tunnels, and buildings were built and increased their sizes by orders of magnitude. It allowed fast, intricate, manufacturing machinery to be made such as that used for making textiles, for agriculture, for printing, and for manufacturing objects. The power levels of steam and gas engines were increased ten-thousand-fold when it became available. This revolutionized transportation on land through railways, and at sea through steam-powered ships. The unprecedented strength, ductility, and economy of steel affected every aspect of the way the world functioned. Unfortunately, it also led to increased destruction in war.

More recently society has been impacted in a major way by the inexpensive air travel that is a direct result of the remarkable strength of nickel–aluminum alloys at high temperatures. For the future, the high strength of pure silicon crystals is being exploited in micro-electronic devices, as well as the strength combined with the optical properties of oxides for optical and materials processing systems. So the importance of the strength of materials to the structure and activities of the world seems clear.

The essence of "solidity" is the ability of a solid to resist shearing forces that try to change its shape. Of all known materials, diamond possesses this ability in the highest degree, and therefore may be considered to be the most "solid" of all substances. In contrast, pure lead is very deficient in this ability. It is much more like a very viscous liquid than a true solid, and if pure, slowly creeps under its own weight. Why the difference? An answer can only be given in terms of the quantum mechanics governing atomic physics. No adequate answer is given by classical continuum mechanics.

The nature of the strength of solids is much more complex than it initially may appear to be. Strength cannot be characterized by a single measurement because, for example, the

1

failure of a filament in simple tension is quite different from the failure of a tubular shaft in torsion. Also, in compression, failure may occur by buckling, but this does not happen in tension. Thus "strength" is a group of properties, not a single one. If forces are applied to a solid object, there are various ways in which it may fail to support them. It may not have enough elastic stiffness. For example, helicopter rotor blades must be stiff enough elastically not to sag (deflect) more than a moderate amount. They must neither droop to the ground when the helicopter is stationary, nor bend upward in flight so much that they cease to generate adequate lift. One of the first challenges faced by the Wright brothers was how to keep the wings of their gliders from deflecting too much.

Although an object may have enough elastic stiffness, it may not have enough resistance to plastic flow. That is, its yield stress may be too low. For example, copper tubing is limited in the amount of water pressure it can sustain by its yield stress. Or, the blades of a high speed turbine may plastically flow slowly (creep), so the turbine fails prematurely. Successful aircraft turbines could not be built in the early part of the twentieth century because of this problem.

A material may be elastically stiff, and resistant to plastic flow, but still fallible if it has low resistance to crack propagation; that is, it is too brittle. Window glass is a classic example. Its chemical bonds are very strong combinations of silicon and oxygen, but cracks propagate through it easily, so its practical strength is rather low. At high temperatures, some metals are elastically stiff, and flow (creep) only slowly, but they rupture because small voids open up in them. They are said to suffer from low stress-rupture strength; or they are said to be "hot short".

In broad terms, resistance to these three failure modes defines what is meant by strength, but there are several complications. One is that the geometric configuration of a material together with the pattern of forces that are applied to it affect how successful it is in avoiding the deformations leading to failure. Some standard configurations are: columns, arches, cables, beams, shafts, shells, and frames (trusses). Each of these has its own characteristics, and behaves differently in tension versus compression versus torsion.

Temperature, time, and chemical reactivity all have large effects on strength. Materials become less solid-like as the temperature rises, until they melt or decompose. Time periods of interest range from the geological ($\sim 10^{15}$ s) to electronic response times ($\sim 10^{-14}$ s), a spread of nearly 30 orders of magnitude! Chemical reactivity may stabilize strength by forming a protective coating, or destabilize it by accelerating crack growth.

Given the large behavioral domain associated with strength, it is not realistic to expect to be able to develop a comprehensive quantitative theory of the behavior in terms of its atomic constituents. At best, the principal ideas regarding the microscopic phenomena that underlie the macroscopic behavior can be described and compared with observations. The set of principal ideas is, in fact, surprisingly small, while the overall complexity is very large. As a result, it is usually easier to measure strength properties than to try to carry out the elaborate calculations that might predict them from physical relationships. However, understanding the physics and chemistry that underlie strength is very useful because it indicates where the limits of the strength properties lie, and it suggests steps that can be taken to improve the practical properties up to their limits. It also indicates how strength properties are related to other physical properties.

The behavior of structural systems is complicated as a result of inescapable interactions between an elastic macroscopic structure (it applies tractions), the elastic response of a material, and the plastic response of the material (or set of materials). Since they are both linear, the first two can usually be combined into one linear response, but plasticity is inherently non-linear. This results in very few situations for which algebraic solutions can be found, and often makes numerical analysis difficult, as well as unreliable. An added difficulty is that there is no equation of state for plastic deformation. Therefore, details of the mechanical history of a material must be taken into account in projecting its future behavior.

A key part of the description of strength is to relate the two kinds of elastic strain (change of volume at constant shape, and change of shape at constant volume) to fundamental physical behavior, and to understand how and why the stiffnesses vary systematically with chemical composition, that is, with positions in the Periodic Table of the Elements, and with various combinations of the elements. The explanation begins with the theory of cohesion which is based on the structures of atoms and the theoretical behavior of electrons.

If solids were perfectly uniform in their structure it would be a relatively simple matter to define their resistance to shearing forces. A single functional relation between stress and strain could be defined for each solid that would depend only on the strength of its atomic bonds and its crystal structure. However, most solids are not perfectly uniform in structure, but contain a heterogeneous host of internal defects. Their strengths are so sensitive to the presence of these defects that the definition of strength acquires great complexity. This structure sensitivity plus the behavioral complexity of electrons in polyelectron atoms is why no comprehensive theory can be expected.

One defect to which all solids are vulnerable is the crack. By acting as a highly efficient lever, a crack allows relatively small applied forces to break the internal atomic bonds of a solid, and thereby reduce its dilatational and shear resistances to zero.

Crystalline solids are also very vulnerable to dislocations. These defects allow shearing to cause shape changes with great ease because they can move and multiply readily, particularly in metals. Other defects which must be considered are atomic vacancies and interstitials, foreign atoms, phonons, excitons, and electrical (color) centers.

An engineering property such as fatigue strength, yield strength, or fracture strength is determined not only by atomic bond strengths, but also by the many defects and their interactions. This makes the subject rich in detail, but often confusing, and sometimes treacherous for those seeking to develop it, or use it.

Time is a factor of primary importance in mechanical behavior. Indeed most of the distinction that is ordinarily made between liquids and solids is based on a shift of time scale. Slowly strained solids behave much like liquids, and liquids subjected to very high strain rates (above about 10^6 s^{-1}) behave much like solids.

In most cases, the resistance to plastic flow is determined by average dislocation mobilities, and these vary enormously (fluxes ranging from zero to approximately 10^{12}/cm^2)

depending on the primary, secondary, and tertiary chemical structure of a material. Not only classical microstructural mechanics is involved, but also quantum-chemical mechanics. Indeed, if electrons were not limited in their behavior by the Pauli Principle of quantum mechanics, it would be a mushy world! It is the localization of electrons required by the Pauli Principle that leads to covalent bonds, and thence to the hard solids of which diamond is the prototype.

Understanding of strength properties in terms of microscopic mechanisms has been relatively slow to develop. In part this has resulted because these properties range so widely over a multitude of variables; and they depend in quite subtle ways on the underlying electronic structures of solids. The range of the properties is approximately the same as that of electrical conductivity for which the ratio of the maximum to the minimum is about 10^{32}. The analogous mechanical transport property is the plastic deformation rate. This has been observed as low as 10^{-16} s^{-1} and, under shock loading conditions, as high as 10^{14} s^{-1}, for a ratio of 10^{30}. Thus the range is similar to that of the electrical resistivity.

It has gradually become clear that detailed knowledge of the nature of the chemical bonding between the atoms of a solid is crucial to an adequate understanding of strength. Chemical bonding is provided by electronic structure. This depends, in turn, on the unique properties of electrons (namely, their charges, masses, wavelengths, phases, and spins).

The sizes of atoms are determined by energies of the electrons in their occupied quantum states. Atomic "shapes" also play a role. These are determined by the angular momenta of the electrons in the occupied energy states. Interactions between the atoms provide cohesion through a rearrangement of the distribution of the valence (bonding) electrons between pairs, and larger clusters of atoms. This electron distribution depends, in addition to other factors, on the "spins" of the electrons which are constrained by the Pauli Exclusion Principle. Therefore, the details of the spacial distributions of the bonding electrons determine the strength properties, not just the average electron energies.

There are four primary properties to be considered: cohesion, elastic stiffness, plastic yield stress, and ductility (fracture). Additionally, there are many secondary properties, but the four primary properties are the crucial ones in the construction of load-bearing structures, and the discussion here will be concerned with relating them to:

(1) the properties of atomic particles (electrons, ion cores, photons, phonons, etc.),
(2) the bonding of atoms (covalent, metallic, ionic, and dispersive),
(3) extrinsic and intrinsic dislocation mobilities,
(4) the behavior of cracks.

It will be assumed that the reader has some knowledge of macroscopic solid mechanics, and therefore with the standard terminology. It will also be assumed that the reader has been introduced to atomic physics, although the pertinent aspects of this will be reviewed. The mathematical level of the discussion will be that of undergraduate physics.

Section II
Elements of solid mechanics

1
Nature of elastic stiffness

The theory of elasticity is a structure of beauty and complexity. No one person sat down and wrote it out. From its first glimmerings in the mind of Galileo (ca. 1638) to the settlement of the question of the minimum number of coefficients required to specify a general elastic response took about 250 years. The latter work was done by Voigt (ca. 1888). Even then very little was known about the underlying factors that determine the coefficients; that is, the chemical properties that determine how resistant a material is to changes in its volume (the bulk modulus), and changes in its shape (the shear moduli). Even now the theory of shear moduli is only partially satisfactory.

Given the ongoing controversy regarding which comes first, practice or science, it is of interest to note that the early history of the theory of elasticity was motivated by the practical interests of those who made the important early advances. Leonardo da Vinci was interested in the design of arches. Galileo was concerned with naval architecture. Hooke wanted to make better clock springs. And Mariotte needed to build effective water piping to supply the palace at Versailles.

Complexity in the theory of elasticity is unavoidable because elastic behavior is intrinsically three-dimensional. Furthermore, most materials are not structurally isotropic; they have textures. Something as simple as a wooden post with a square cross-section looks different, in general, along each of the perpendiculars to its faces. These structural differences translate into different elastic stiffnesses in the three perpendicular directions. The situation is further complicated by the need to describe the shear responses in terms of a shear plane, plus a direction on the plane.

If forces are applied to the opposite ends of a slender bar, it will elongate or contract in proportion to the size of the forces (Hooke's Law). The coefficient of response is Young's modulus. However, the bar might also be twisted around its length, or it might be bent around an axis perpendicular to its length. The twisting mode of deformation requires another response coefficient, the shear modulus. If the slenderness of the bar is reduced to the limit of a line, both of its elastic moduli become meaningless. They cannot exist in the absence of finite atoms, and the bonds between the atoms which are formed by electrons that behave according to quantum mechanics.

Consider a square mesh of wires soldered together at the intersections of the wires (a wire screen). Suppose a square piece is cut from the mesh with the edges of the square parallel to the wires. If forces are applied parallel to the edges of the square, the response will be stiff, whereas if forces are applied parallel to one of the diagonals of the square, the response will be relatively soft. Thus two distinctly different response coefficients are needed to describe the square's mechanical behavior.

Next imagine a three-dimensional framework of wires in a cubic array soldered together at the nodes of the wires (on a larger scale this would be like the framework of a steel building with all of the girders of the same length). This will have three different response coefficients in general: one for forces applied parallel to the wires, another for forces applied parallel to the diagonals of the faces of the cubes, and the third for forces applied along the diagonals of the cubes (lines connecting the opposite far corners). The three coefficients can be reduced to two by adjusting the design of the nodes; then the structure is said to be elastically isotropic.

More response coefficients will be required if the symmetry of the structure is less than that of a cube. For example, if the framework consists of rectangular parallelepipeds, the number of coefficients increases to nine (orthorhombic symmetry). Then if the number of different lengths is reduced from three to two (tetragonal symmetry), the number of coefficients drops from nine to six; and if the number of different lengths is further reduced to one, only three coefficients are needed (cubic symmetry) as already indicated.

Clearly, a standardized framework is necessary within which the elastic coefficients can be defined. This is provided by tensor calculus which, as the name suggests, was devised for this purpose. It is necessarily more complex than vector calculus because the elastic state of a solid requires both tensions (or compressions) and shears to describe it.

The first step is to define what is meant by *stress* and by *strain*. This will include a definition of the notation that is used to distinguish the various possible components of the stress and strain tensors. Then the response coefficients that connect them can be defined.

One of the purposes of this book is to show how the response coefficients (elastic constants) are determined by the chemical and physical constitutions of various solids. This involves the interior geometry of the solid, and its corresponding electronic structure.

Temperature and time both have small and large effects on elastic stiffness depending on the material. At very low temperatures elastic stiffness becomes independent of temperature, but at higher temperatures (near the Debye temperature and above) various *anelastic* effects occur. Then a given material has two stiffnesses: one for fast loading when there is too little time for anelastic relaxation (called the unrelaxed modulus), and the other for slow loading which allows relaxation to occur (called the relaxed modulus). The difference between these two stiffnesses is small for solids in which the atoms are densely packed. A much larger effect is found in less dense materials such as elastomers (rubber-like materials). In these, the molecules tend to curl up into coils which have high entropy because they can be formed in many different ways. When applied tractions stretch them, there are fewer ways for them to coil so their entropies decrease. This increases their free energies, so they resist stretching, but relatively weakly. They are sometimes said to have "entropic elasticity", as contrasted with the more usual "enthalpic elasticity".

The elastic response coefficients are the most fundamental of all of the properties of solids, and the most important sub-set of them is the shear coefficients. If these were not sufficiently large, all matter would be liquid-like. There would be no aeronautical, civil, or mechanical engineering. Furthermore, modern micro-electronics, as well as opto-electronics would not be possible. The elastic stiffnesses set limits on how strong materials can be, how slowly geological processes occur, and how natural structures respond to wind and rain. This is why the scientific study that began with Galileo continues today.

Imagine a world in which everything has the same elastic stiffness. If all the bulk stiff-nesses (the resistance to volume changes) were the same, nails could not be driven into wood, and plows could not turn earth. Or, suppose that the stiffness of aluminum were one-fifth as large as it actually is. Then the wing tips of large aircraft would drag on the ground because the elastic deflections would be so large. It is for reasons like these that the elastic properties of solids have great engineering significance, and why the theory of elasticity played such an important role in the histories of both engineering and physics.

The architecture of the theory of elasticity is now considered to be applied mathematics, but once was in the mainstream of the development of calculus and differential equations, as well as physics. For a long time (centuries), the elastic properties were coefficients to be measured and tabulated. Their relationship to the properties of atomic particles, and to one another, awaited the development of quantum mechanics. Although there remain some aspects of the theory that are not entirely satisfactory, the progress that has been made toward a general theory is impressive.

All forms of matter (gases, liquids, solids, plasmas, etc.) resist changes of volume, and the amount of this resistance is measured by means of the bulk modulus. Its inverse is the compressibility. Solids are defined by their shear stiffness moduli. These have inverses called shear compliances. Since the shear response is difficult to separate from the volumetric response, the overall description of elastic behavior is complicated.

The primary factor determining elastic stiffness is chemical constitution because it de-termines the internal bonding. Broadly, there are four kinds of bonding: covalent, ionic, metallic, and molecular. Each has its idiosyncrasies. The stiffest bonds are of covalent character, while the least stiff are molecular.

In addition to their shear stiffnesses, solids have another special feature. They can be either perfect, nearly perfect, or imperfect, in terms of their structural geometry. An ideal, or perfect, solid has a specific crystal structure, and each site of the crystal structure is occupied by a specific atomic species. However, virtually all solids contain defects, including thermal vibrations, vacancies, interstitials, impurity atoms, dislocations, stacking faults, domain boundaries, and grain boundaries. These affect the elastic stiffness, particularly the shear stiffness.

The time dependence of elasticity, or *anelastic* response, results from a variety of effects. Some of these are the thermo-elastic effect, the hopping of carbon atoms in iron (Snoek effect), and the stress-induced ordering of atomic pairs in some alloys. The anelasticity of elastomers (rubber) is a much larger effect.

Changes of shape (strains) can be induced by fields other than stresses. Electric fields cause electrostriction, or piezoelectric strains; and magnetic fields cause magnetostriction.

An aim here is to describe the connections of the various response coefficients to chemical constitution. This is unlike many books on strength which describe the field variables, leaving the elastic stiffnesses as coefficients to be measured and tabulated.

2

Generalized stress

Stress is a generalization of the concept of pressure. The latter consists of a unit of force applied perpendicular to a unit of area. However, in the case of stress, the force can be applied at any angle relative to the unit area. Thus, whereas pressure is a scalar that does not act in any special direction, two vectors are needed to define a general stress: one that indicates the direction of the force, and the other that gives the orientation of the area on which the force acts. Furthermore, in the case of solids, the orientation of the solid must be defined, except when the solid is isotropic. Stresses applied to surfaces (force per unit area) are called *tractions*. They can be resolved into two components: one parallel, and the other perpendicular to the unit surface areas on which they act. The orientation of the area is given by a unit vector lying perpendicular to it. For the force component that lies parallel to the unit surface area a *shear* traction is produced, and for the force component that lies perpendicular to the unit surface, a *normal* traction is produced.

Tractions tend to distort a solid material, either causing its volume to change, but not its shape, or causing its shape to change at constant volume (small changes), or both. That is, they create *dilatational strain* in the material, or *shear strain*, or both. Hooke's Law states that the amount of strain created is proportional to the amount of traction applied in the elastic regime. A corollary is that the strain disappears if the traction is removed. This was first clearly demonstrated and understood by Robert Hooke (Timoshenko, 1983). He was interested in designing springs to drive clockworks. This led him to measure the elastic behavior of wires in simple tension, bending, and torsion.

The existence of recoverable strains implies the existence of internal forces that balance the externally applied tractions. Expressed as forces per unit area, these are the *stresses*. It is assumed that the distribution of forces is continuous, and that the material is continuous and singly connected (an exception occurs when the material contains a *dislocation* so the material becomes doubly connected). Therefore, internal surfaces are continuous. It is also assumed that no "body forces" are acting, which is not always the case. Body forces may arise from inertial effects (angular or linear accelerations), gravitation, magnetic fields, electric fields, and so on.

The responses of materials to stresses were first measured in a rudimentary way by Leonardo da Vinci, ca. AD 1500. Then, Galileo reported the results of systematic studies in his book *Two New Sciences* (mechanics and motion) in 1638. Nearly 200 years passed before

Augustin Cauchy in 1822, and Barré de Saint-Venant in 1845, developed a satisfactory mathematical definition of stress (Timoshenko, 1983).

The stress on an infinitesimal element of a plane within a strained elastic material is defined, according to Saint-Venant, as "the resultant of all the actions (forces) of the molecules situated on one side of the plane upon the molecules on the other side; the directions of which intersect the element. The magnitude of the stress is obtained by dividing the resultant by the area of the element." This definition had to await the invention of the calculus to become useful.

In most cases a Cartesian coordinate system is used. Then, if the solid is fibrous, one of the coordinate axes (usually the z-axis) is taken to be parallel to the fiber axis. If the solid is a textured plate, such as a piece of rolled steel, one coordinate axis is taken parallel to the rolling direction and another perpendicular to the plane of the plate. If a crystal is being described, its crystallographic axes must be taken parallel to specific coordinate axes. For the most common cases of cubic, tetragonal, and orthorhombic symmetry, the crystallographic axes **a**, **b**, **c** are orthogonal so they are simply taken to be parallel to the orthogonal coordinate axes **1**, **2**, **3**.

As mentioned already, the local force can arise in various ways:

(1) externally applied tractions,
(2) interactions with electric, magnetic, or gravitational fields,
(3) local accelerations associated with wave propagation,
(4) non-uniform temperature distributions,
(5) internal distortions associated with dislocations, disclinations, and other defects.

2.1 Specification of a plane

The most convenient way to specify a plane is through the use of a unit normal vector, that is, a vector of unit length lying perpendicular to the plane. Its projections onto a Cartesian coordinate system equal its direction cosines (the cosines between the vector and the coordinate axis vectors). This is illustrated in Figure 2.1 where the coordinate system is given by the three orthogonal vectors **1**, **2**, **3**, and **r** is a unit vector parallel to any direction starting at the origin. The direction cosines of **r** are c_{ir} where $i = 1, 2, 3$. The sum of their squares equals unity:

$$(c_{1r})^2 + (c_{2r})^2 + (c_{3r})^2 = 1 \tag{2.1}$$

or, using the notation that summation is to occur over any repeated subscript:

$$(c_{ir})^2 = (c_{ir})(c_{ir}) = 1 \tag{2.2}$$

The unit vector **r** can also be used to designate the plane that is perpendicular to it. If another plane is designated by **s**, then the cosine of the angle between the two planes is:

$$c_{rs} = c_{ir}c_{is} \tag{2.3}$$

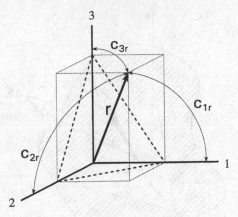

Figure 2.1 Definition of an element of a plane by means of a normal vector **r** with the direction cosines of the angles: c_{1r}, c_{2r}, and c_{3r}.

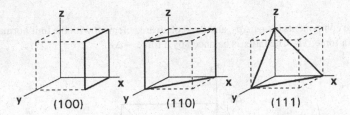

Figure 2.2 The three most important planes in the cubic symmetry system with their Miller indices: (100), (110), and (111).

Another convenient way to designate a plane is through its *Miller indices*. These are found as follows. In a crystallographic coordinate system, **a**, **b**, and **c** determine the integral intercept of the plane along each of the coordinates. The reciprocal of each intercept is formed. Then the reciprocals are multiplied by the lowest common denominator (if there are no fractional reciprocals, this step is unnecessary). Note that the reciprocal of infinity is zero. Figure 2.2 shows the three most important planes in the cubic symmetry system together with their Miller indices.

The Miller indices of a direction are those of the plane that lies perpendicular to the direction, with the indices put into square brackets. Thus the direction [*hkl*] lies perpendicular to the plane (*hkl*). The group of *any* (*hkl*) planes is written {*hkl*}, and the group of *any* [*hkl*] directions is ⟨*hkl*⟩.

2.2 Resolution of an area element

In three dimensions, a small element of area whose unit normal vector is Δ**a** can be resolved into three components. Each of these may be taken to be perpendicular to one of the three coordinate axes, as in Figure 2.3. Therefore, Δ**a** has three components, $\Delta a_j (j = 1, 2, 3)$.

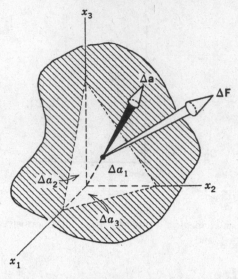

Figure 2.3 The components (Δa_1, Δa_2, and Δa_3) of an area element with the unit normal vector, $\Delta\mathbf{a}$, acted upon by a force $\Delta\mathbf{F}$. The latter is balanced by a force $-\Delta\mathbf{F}$.

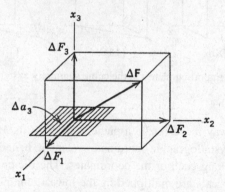

Figure 2.4 Three components of the force $\Delta\mathbf{F}$ acting on one of the area elements of Figure 2.3, Δa_3. One of the components, ΔF_3, acts normal to the area, while the other two, ΔF_1 and ΔF_2, act tangential to the area.

2.3 Resolution of a force element

The force element $\Delta\mathbf{F}$ which acts on $\Delta\mathbf{a}$ in Figure 2.3 can be resolved into three components each parallel to one of the coordinate axes (Figure 2.4). These components are designated $\Delta F_i (i = 1, 2, 3)$.

2.4 Definition of the local state of stress

Each of the three components of $\Delta\mathbf{F}$ acts on an area component such as Δa_3. Thus three forces per unit area, i.e. stresses (σ_{ij}), act on Δa_3. They have the forms σ_{i3}:

$$\frac{\Delta F_1}{\Delta a_3} = \sigma_{13} \qquad \frac{\Delta F_2}{\Delta a_3} = \sigma_{23} \qquad \frac{\Delta F_3}{\Delta a_3} = \sigma_{33}$$

Similar sets of stresses act on Δa_1 and Δa_2. One of each set acts perpendicular to the area element, and two act tangential to it as illustrated in Figure 2.4. The first subscript in each case indicates the direction of the force, and the second indicates the direction of the perpendicular to the area (the area vector).

In the same way, three other stress components (σ_{12}, σ_{22}, and σ_{32}) can describe the forces per unit area exerted on Δa_2. Similarly, the area element Δa_1 has three stress components (σ_{11}, σ_{21}, and σ_{31}) acting on it. In all, nine stress components act on the orthogonal faces of the cubic element, plus nine balancing stresses of opposite sign. The nine stresses form the components of the matrix that is the second-order *stress tensor*:

$$\sigma_{ij} = \begin{array}{ccc} \sigma_{11} & \sigma_{12} & \sigma_{13} \\ \sigma_{21} & \sigma_{22} & \sigma_{23} \\ \sigma_{31} & \sigma_{32} & \sigma_{33} \end{array} \quad i, j = 1, 2, 3 \qquad (2.4)$$

Notice that the stress tensor connects two vectors, each of which has three components. One is a force vector, and the other is the direction cosines of a unit vector that lies normal to a plane of unit area. Thus, as a mathematical object, it is compact, but it can describe complex mechanical states of materials (Sines, 1969). Tensors have the property that, although the individual values of the components will change if the coordinate system is changed, the tensor as a whole will not be affected. The plausibility of this is based on considering either of the component vectors. The independence of a vector from the particular set of coordinates that is used to describe it can be readily visualized. A proof for tensors can be found in texts on tensor analysis, or crystallography (Nye, 1957).

The name "stress tensor" for Equation (2.4) was coined by Waldemar Voigt. He chose the word "tensor" to suggest tensile stresses. His monograph, *Lehrbuch der Kristallphysik*, is a classic of the physics literature (Voigt, 1910).

The forces acting on one cross-section of a volume element due to σ_{ij} are sketched in Figure 2.5. It can be seen that the shear couple formed by σ_{12} must be balanced by the

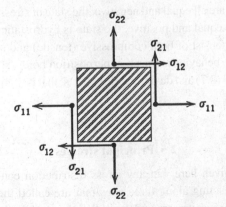

Figure 2.5 Static equilibrium of the stresses acting on a block of material. Note that the normal stresses are equal and opposite: $|\sigma_{11}| = -|\sigma_{11}|$, and $|\sigma_{22}| = -|\sigma_{22}|$. Also, the couple formed by σ_{12} is balanced by the σ_{21} couple. Thus, $\sigma_{ij} = \sigma_{ji}$.

couple of σ_{21} in order to have static equilibrium. Therefore:

$$\sigma_{ij} = \sigma_{ji} \qquad i, j = 1, 2, 3 \tag{2.5}$$

and the stress tensor takes the symmetric form:

$$\sigma_{ij} = \begin{matrix} \sigma_{11} & \sigma_{12} & \sigma_{13} \\ \sigma_{12} & \sigma_{22} & \sigma_{23} \\ \sigma_{13} & \sigma_{23} & \sigma_{33} \end{matrix} \tag{2.6}$$

where the diagonal components with $i = j$ describe tensions and compressions, while the off-diagonal components with $i \neq j$ describe shears.

Usually, a state of stress imposes both dilatational and shear forces on a material. Since it is easier to visualize these separately, the complete tensor is often resolved into two parts: a *spherical* part and a *deviator* part.

The spherical part is constructed from the common factor among σ_{11}, σ_{22}, and σ_{33}. This is the pressure, $\pm P$, and the spherical tensor is:

$$\sigma_{ij}^S = \begin{matrix} P & 0 & 0 \\ 0 & P & 0 \\ 0 & 0 & P \end{matrix} \tag{2.7}$$

while the deviator, or shear part is:

$$\sigma_{ij}^D = \begin{matrix} [2\sigma_{11} - (\sigma_{22} + \sigma_{33})]/3 & \sigma_{12} & \sigma_{13} \\ \sigma_{12} & [2\sigma_{22} - (\sigma_{11} + \sigma_{33})]/3 & \sigma_{23} \\ \sigma_{13} & \sigma_{23} & [2\sigma_{33} - (\sigma_{11} + \sigma_{33})]/3 \end{matrix} \tag{2.8}$$

An applied pressure, P, is equal to the average of the diagonal terms of the stress tensor:

$$P = \pm(\sigma_{11} + \sigma_{22} + \sigma_{33})/3 \tag{2.9}$$

If the stress components are all equal and negative, the state of stress is said to be hydrostatic compression. If they are equal and positive, the state is hydrostatic tension.

Most states of stress consist of both compressive (tensile) and shear stresses, but in the elastic regime where the behavior is linear, superposition holds, so any stress state can be factored into its spherical (2.7) and deviatoric (2.8) parts, that is, into a pressure (hydrostatic) part and a shear part.

2.5 Principal stresses

The proof will not be given here, but any stress distribution can be resolved into three orthogonal components acting along three axes that are called the *principal* axes. These three components are normal stresses and are called the *principal stresses*: σ_1, σ_2, and σ_3. The planes that are perpendicular to the principal stresses have no shear stresses acting on them. The stress tensor becomes:

$$\sigma_{ij} = \begin{matrix} \sigma_1 & 0 & 0 \\ 0 & \sigma_2 & 0 \\ 0 & 0 & \sigma_3 \end{matrix} \tag{2.10}$$

where the three principal stresses act in three orthogonal *principal directions*. If the three principal stresses are equal to one another, then there are no shear stresses on any planes, and the state of stress is said to be *hydrostatic*.

If one of the principal stresses is zero, and the other two are equal, the stress state is said to be *plane biaxial*. An example is a membrane, such as the center of a taut trampoline. If the two principal stresses are equal in magnitude, but opposite in direction (the third being zero), the stress state is *pure plane shear*. This is nearly realized in the wall of a thin-walled tube that is loaded in torsion, but it is difficult to realize it otherwise.

If two of the principal stresses are zero, the state of stress is uniaxial, in tension or compression. In this case there is no shear on the plane perpendicular to the single principal stress, but there is finite shear on all other planes. The shear stress is maximal on the planes that lie at 45° to the principal stress axis. Its sign depends on the sign of the principal stress.

The stress tensor is called a *field tensor* since it can have any orientation relative to a piece of material. In this respect, it differs from a *matter tensor* which describes a property and must have a specific orientation relative to any anisotropies that a piece of material may have. In particular, for crystals, the orientation of a matter tensor must be defined relative to the crystallographic axes. These axes must, in turn, be defined relative to a set of Cartesian, or other, coordinate axes. This requires conventions to be defined if numerical values of the properties are to have the same meanings for everyone who works on the subject. These conventions have been established by the IEEE (Institute of Electrical and Electronic Engineers) for the properties of crystals and are given in the book by Nye (1957).

The next step is to describe a state of strain (Chapter 3).

References

Nye, J.F. (1957). *Physical Properties of Crystals*. Oxford: Clarendon Press.
Sines, G. (1969). *Elasticity and Strength*. Boston, MA: Allyn and Bacon.
Timoshenko, S.P. (1983). *History of the Strength of Materials*. New York: McGraw-Hill, 1953.
    Republished by Dover Publications, New York.
Voigt, W. (1910). *Lehrbuch der Kristallphysik*. Leipzig: Teubner.

3

Generalized strain

External forces (tractions) applied to a solid may translate it, rotate it, or induce stresses in it which strain it. Stresses are induced if the forces create a non-uniform field of displacements u_i within the solid. If the displacement field is uniform, translation and/or rotation of the solid may occur but no strains (or stresses) will be induced. Thus the displacements must be a function of position, $u_i(x, y, z)$, if there are to be strains or local rotations. Since displacements only exist relative to some initial reference state, they are relative quantities. This makes the analysis of strain rather subtle, requiring unusually precise definitions.

If the response of a structural material is to be elastic (i.e., reversible) the displacement gradients, and local rotations, in it must be small.[1] Therefore, they can be described locally by a linear vector field: $u_i = \delta_{ij} r_j$ where the repetition of the index j means summation over j, so for example $u_1 = \delta_{11} r_1 + \delta_{12} r_2 + \delta_{13} r_3$.

For a local position, the δ_{ij} are the slopes in various directions, so:

$$\delta_{ij} = \partial u_i / \partial r_j \qquad (3.1)$$

If u_i is a constant (uniform), $\delta_{ij} = 0$ and there may be translation, but no strain as stated previously.

The vector displacements of neighboring points cannot be completely arbitrary. If they were, holes in the material might open up, or more than one particle might try to occupy the same space. Therefore, there are constraints known as *compatibility relations* that will be described later.

Figure 3.1 may clarify this discussion. In an x, y, z coordinate system it shows a point, P, located by a vector, $\mathbf{r}_o$. It is displaced to a point, P', located by $\mathbf{r}'$. Its displacement is: $\mathbf{u} = \mathbf{r}' - \mathbf{r}_o$. Notice that (in general) this does two things: it changes the length of $\mathbf{r}$, and it changes the angular position of $\mathbf{r}$. Thus it causes an extension (or compression), and a shear, or a rotation. In terms of the gradients, δ_{ij}, the extensions (compressions) correspond to $i = j$, while the shears and rotations are described by the coefficients with $i \neq j$.

---

[1] Strictly speaking, the strain must also be imposed slowly enough for the material to maintain a constant temperature, so the modulus is *relaxed*. Otherwise, through the thermo-elastic effect, the temperature will be changed a small amount by the strain. In compression it will rise, and in extension it will fall (like the behavior of a gas). The effect is small, and will be neglected in this book. Elastomers behave differently, and will be discussed later.

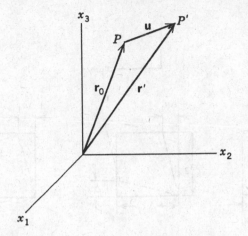

Figure 3.1 Definition of a displacement vector.

For the simple case of a thin wire being stretched along the 1-axis, $u_1 = \delta_{11} r_1$, so the displacement increases from the origin as r_1 increases. The nominal tensile deformation is δ_{11}. This is a component of the *deformation tensor* which connects position and displacement:

$$u_i = D_{ij} x_j \tag{3.2}$$

$$D_{ij} = \begin{matrix} D_{11} & D_{12} & D_{13} \\ D_{21} & D_{22} & D_{23} \\ D_{31} & D_{32} & D_{33} \end{matrix} \quad i, j = 1, 2, 3 \tag{3.3}$$

which can be written out as:

$$\begin{aligned} u_1 &= u_1^0 + (\partial u_1/\partial x_1)x_1 + (\partial u_1/\partial x_2)x_2 + (\partial u_1/\partial x_3)x_3 \\ u_2 &= u_2^0 + (\partial u_2/\partial x_1)x_1 + (\partial u_2/\partial x_2)x_2 + (\partial u_2/\partial x_3)x_3 \\ u_3 &= u_3^0 + (\partial u_3/\partial x_1)x_1 + (\partial u_3/\partial x_2)x_3 + (\partial u_3/\partial x_3)x_3 \end{aligned} \tag{3.4}$$

where the $(\partial u_i/\partial x_j) = D_{ij}$. If the deformation is homogeneous, it is independent of the origin of the coordinates u_i^0 and this vector can be neglected. For small deformations, it is reasonable to assume they are homogeneous, but if they are finite, changes of the position of the origin must be considered.

For a constrained volume element, all of the components of D_{ij} lead to internal distortions and therefore to changes of the internal energy. However, some of the components describe a uniform rotation of the whole element, so if the element is unconstrained, the atomic particles within it maintain their positions relative to one another, so no change of internal energy occurs. Because of this, although it leads to some confusion, it is customary to factor the deformation tensor into two parts: a *rotation* tensor ω_{ij} and a *strain* tensor ϵ_{ij}. The forms

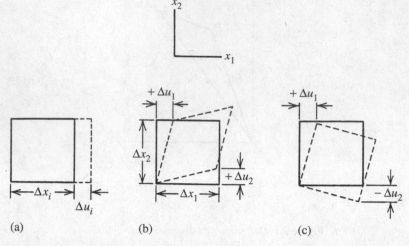

Figure 3.2 Illustration of deformations: (a) stretching for Δu_i positive, and compression for Δu_i negative; (b) shear; (c) rotation.

of these are as follows:

$$\omega_{ij} = (1/2)[(\partial u_j/\partial x_i) - (\partial u_i/\partial x_j)] = \begin{matrix} 0 & -\omega_{21} & +\omega_{31} \\ +\omega_{12} & 0 & -\omega_{32} \\ -\omega_{13} & +\omega_{23} & 0 \end{matrix} \quad (3.5)$$

$$\epsilon_{ij} = (1/2)[(\partial u_j/\partial x_i) + (\partial u_i/\partial x_j)] = \begin{matrix} \epsilon_{11} & \epsilon_{12} & \epsilon_{13} \\ \epsilon_{21} & \epsilon_{22} & \epsilon_{23} \\ \epsilon_{31} & \epsilon_{32} & \epsilon_{33} \end{matrix} \quad (3.6)$$

The strain tensor can be further factored into two parts, the first describing dilatations, and the second describing shears:

$$\epsilon_{ij} = \begin{matrix} \epsilon_{11} & 0 & 0 \\ 0 & \epsilon_{22} & 0 \\ 0 & 0 & \epsilon_{33} \end{matrix} + \begin{matrix} 0 & \epsilon_{12} & \epsilon_{13} \\ \epsilon_{21} & 0 & \epsilon_{23} \\ \epsilon_{31} & \epsilon_{32} & 0 \end{matrix} \quad (3.7)$$

Figure 3.2 illustrates the three principal deformations.[2] On the left, stretching is illustrated where the change of length Δu_i is parallel to the length itself Δx_i. In the middle, shearing is shown where the change of displacement is perpendicular to the length parameter, and the sum of $(\Delta u_j/\Delta x_i)$ and $(\Delta u_i/\Delta x_j)$ defines the shear angle. On the right, the changes of displacement are again perpendicular to the length parameter, but now their difference defines the rotation angle.

The shear strains in Figure 3.2 are symmetric, so:

$$\epsilon_{ij} = \epsilon_{ji}$$

[2] Note that tensorial strains, ϵ, are only half as large as "engineering strain components", designated e to distinguish them.

while the rotations are anti-symmetric:

$$\omega_{ij} = -\omega_{ji}$$

Furthermore, the *shear angle* θ_{12} is:

$$\theta_{12} = 2\epsilon_{12} = 2\epsilon_{21}$$

so it is important in doing numerical calculations to distinguish between a strain defined by the shear angle, and one defined by the components of the strain tensor. The symmetric strain tensor is:

$$\epsilon_{ij} = \begin{matrix} \epsilon_{11} & \epsilon_{12} & \epsilon_{13} \\ \epsilon_{12} & \epsilon_{22} & \epsilon_{23} \\ \epsilon_{13} & \epsilon_{23} & \epsilon_{33} \end{matrix} \qquad i, j = 1, 2, 3 \qquad (3.8)$$

This has only six independent terms. When $i = j$, the strain is said to be a *normal* strain; it is a *shear* strain when $i \neq j$.

For an isotropic change of volume:

$$\Delta V/V = \epsilon_{11} = \epsilon_{22} = \epsilon_{33} = (\epsilon_{11} + \epsilon_{22} + \epsilon_{33})/3 \qquad (3.9)$$

which is called a *dilatation*, or, if it is negative, the state of strain is called *hydrostatic compression*. In this case the shape does not change ($\epsilon_{ij} = 0$, $i \neq j$). However, if $\epsilon_{ii} = 0$, whereas some of the $\epsilon_{ij} \neq 0$, the shape changes at constant volume (for small strains). Such shape changes are called *distortions*.

Dilatations cause changes in the *distances* between pairs of atomic particles. Distortions change the *angles* between three or more of them. Rotations do not cause such changes. Therefore, it is commonly assumed that distortions and dilatations cause changes in the internal energies of solids while rotations do not. However, if a solid contains long-range internal force fields (such as those associated with ferromagnetism and electric polarization) then locally rotated material may interact with the non-rotated matrix, and thereby change the internal energy.

If the strain distribution in a material is constant, or if it varies linearly with position in a material, then the system of internal strains and surface tractions is statically determinate. Otherwise, the system is statically indeterminate, but it must satisfy the *compatibility relations*. If these are satisfied, together with the equilibrium conditions, and the boundary conditions, then the distribution of stresses and strains that has been deduced is unique (unambiguous) (Timoshenko, 1934).

Compatibility is particularly important when plastic deformation is involved because the deformation tends neither to be constant, nor to vary linearly with position. It is common for there to be *incompatibility* at grain boundaries which causes cracking.

Reference

Timoshenko, S. (1934). *Theory of Elasticity*. New York: McGraw-Hill.

4

Elastic coefficients

The stress and strain tensors that have just been defined now need to be connected. According to the experiments of Robert Hooke (1678), the components of the stress and strain tensors are linearly proportional to one another (for small strains). Since the variables are tensors, the coefficients of proportionality must belong to another tensor. This is called the *elasticity tensor*. It has two variants depending on whether stress or strain is taken to be the independent variable. If stress is the independent variable with strain proportional to it, then the coefficients of the elasticity tensor are called *elastic compliances*. For the converse case, where the strain is applied and stress is the response, the coefficients are called *elastic stiffnesses*.

Stress and strain are second-rank tensors (vectors are first-rank tensors, and scalars are zero rank), while the tensor of elastic coefficients is a fourth-rank tensor (note that the rank equals the number of subscripts following its coefficients). The tensor equation that expresses Hooke's Law in terms of the elastic stiffnesses is as follows:

$$\sigma_{ij} = \sum_{kl} C_{ijkl}\epsilon_{kl} \qquad \text{where } i, j, k, l = 1, 2, 3 \qquad (4.1)$$

As the resistance of a solid to elastic deformation increases, so do the magnitudes of the C_{ijkl} coefficients. Since this is a linear equation, it can be inverted to obtain the elastic compliance tensor S_{ijkl} which relates strains to stresses:

$$\epsilon_{ij} = \sum_{kl} S_{ijkl}\sigma_{kl} \qquad (4.2)$$

For quasi-static measurements, Equation (4.2) is the more useful form of Hooke's Law because it is easier to apply a single stress component and then measure all of the resulting strains than it is to apply a single strain component and attempt to measure the resulting stresses. Once the compliances are known, the tensor can be inverted to obtain the stiffnesses (Nye, 1957). However, for dynamic measurements (wave propagation) it is easier to apply Equation (4.1). In either case, once one of the elasticity tensors is known, the other one can be determined by inversion. The inversion equation can be written in terms of the Kronecker delta function: $\delta_{km} = 1$ if $k = m$, and $\delta_{km} = 0$ if $k \neq m$. Thus the compliance and stiffness tensors are related by:

$$C_{ijkl} S_{ijmn} = \delta_{kn} \delta_{lm} \tag{4.3}$$

The factors that determine the magnitudes of the coefficients in these tensors are one of the concerns of this book. Continuum solid mechanics cannot account for their magnitudes, so it is incomplete. A theory of the coefficients requires quantum mechanics to show how atomic bonding energies and bonding forces arise.

The elastic response coefficients vary considerably from one material to another. Graphite on the high end has an in-plane elastic stiffness of about 690 GPa, while elastomers on the low end have stiffnesses of about 10^{-3} GPa (the range is about 10^6). Some materials are very nearly isotropic (for example, aluminum and tungsten), while others are highly anisotropic. Examples of the latter are graphite, mica, and talc crystals. Fibrous materials can be even more anisotropic where the stiffness parallel to the fiber axes is very large, while the transverse stiffness is very low.

A complete description of the elastic state of a solid requires knowledge of both the stress and the strain tensors. Each of them has nine components, so the most general elasticity tensor has 81 components, but equilibrium and symmetry conditions reduce this considerably. Since there can be no net torques on an elastic body, the shear stress components must be balanced in pairs, so $\sigma_{ij} = \sigma_{ji}$. This reduces the 9×9 matrix of coefficients to a 6×6 matrix with six diagonal coefficients and 30 off-diagonal ones. Furthermore, since elastic deformation is reversible, elastically strained solids contain strain energy. This is a scalar quantity proportional to the square of the strain which requires the matrix of elastic constants to be symmetric. Therefore, 15 of the off-diagonal coefficients on one side of the diagonal equal 15 on the other side. This leaves $21 (6 + 15)$ independent coefficients in the most general case of a solid with triclinic symmetry (all three crystal axes of different lengths, and three different angles between the axes). Many solids have more symmetry than this, however, so the number of coefficients can be reduced substantially further.

Suppose the solid is isotropic. Then all of the off-diagonal coefficients must equal zero because stresses in one direction cannot be coupled with strains in some other direction if isotropy is to be maintained. Furthermore, the response of the material to a normal stress must not depend on the direction in which it is applied so the three diagonal coefficients that couple normal stresses to extensions (or compressions) must be equal. Similarly, the three diagonal shear constants are also equal. As a result only two coefficients are left. These two are the bulk modulus, B, for isotropic normal (hydrostatic) stresses, and the shear modulus, G, for shear stresses.

In addition to the symmetry elements that all materials have, some materials have internal symmetries. For example a crystal that has cubic symmetry must look the same elastically when it is viewed along the x_1-axis as when it is viewed along the x_2-axis, or the x_3-axis. Therefore, it cannot have more than three independent elastic constants. One of these is chosen to describe stretching (or compressing), and the other two to describe two shear systems.

In general, the crystallographic axes ($\mathbf{a}$, $\mathbf{b}$, $\mathbf{c}$) are not orthogonal, but the elasticity tensors are referred to the orthogonal axes ($\mathbf{x}_1$, $\mathbf{x}_2$, $\mathbf{x}_3$). Therefore, as mentioned in Chapter 2, a convention is needed to link the two sets of axes. A standard set of conventions has been specified by the Institute of Electrical and Electronics Engineers (IEEE). For crystals with monoclinic symmetry, the IEEE convention is not the same as the older convention of Voigt in his *Lehrbuch der Kristallphysik*. Thus some caution is needed.

To demonstrate the effects of crystal symmetries, consider a specific constant in the cubic symmetry system, namely, C_{1123}. It is assumed to be positive. Then, according to Equation (4.1), a strain $+\epsilon_{23}$ causes a stress $+\sigma_{11}$. But, suppose that the crystal is rotated around the $\mathbf{x}_3$-axis, so $+\mathbf{x}_2$ becomes $-\mathbf{x}_2$. Then $+\epsilon_{23}$ nominally becomes $-\epsilon_{23}$, although it does not change physically because the x_2x_3-plane has mirror symmetry. Since $\sigma_{11} = C_{1123}\epsilon_{23}$, this means that σ_{11} is hypothetically negative, but a change of coordinates cannot change the sense of a physical quantity. Hence, it must be concluded that $C_{1123} = 0$ in the cubic system.

Furthermore, in the cubic system there is mirror symmetry about all three planes ($\mathbf{x}_1$, $\mathbf{x}_2$), ($\mathbf{x}_2$, $\mathbf{x}_3$), and ($\mathbf{x}_3$, $\mathbf{x}_1$). Therefore, C_{ijkl} must be invariant for all changes of the signs of $\mathbf{x}_1$, $\mathbf{x}_2$, or $\mathbf{x}_3$. This means that all terms of the form C_{iiij} must be equal to zero. Only terms of the form C_{iijj}, C_{ijij}, or C_{iiii} are non-zero. Also, not all terms of the last form are different. Symmetry requires some equalities between them. A cubic crystal looks the same in the three directions: $\mathbf{x}_1$, $\mathbf{x}_2$, $\mathbf{x}_3$. In other words, its behavior is independent of the transformations:

$$
\begin{aligned}
1 &\rightarrow 2 & 1 &\rightarrow 3 \\
2 &\rightarrow 3 & 2 &\rightarrow 1 \\
3 &\rightarrow 1 & 3 &\rightarrow 2
\end{aligned}
$$

so

$$C_{1111} = C_{2222} = C_{3333}$$

and

$$C_{1122} = C_{3322} = C_{2211} = C_{1133} = C_{3311} = C_{2233}$$

as well as

$$C_{1212} = C_{2323} = C_{1313} = C_{2121} = C_{3232} = C_{3131}$$

It follows that the matrix of coefficients contains only three independent constants, and has the following appearance:

$$
\begin{bmatrix}
C_{1111} & C_{1122} & C_{1122} & 0 & 0 & 0 \\
C_{1122} & C_{1111} & C_{1122} & 0 & 0 & 0 \\
C_{1122} & C_{1122} & C_{1111} & 0 & 0 & 0 \\
0 & 0 & 0 & C_{1212} & 0 & 0 \\
0 & 0 & 0 & 0 & C_{1212} & 0 \\
0 & 0 & 0 & 0 & 0 & C_{1212}
\end{bmatrix}
$$

For an isotropic crystal, or other isotropic material, the constants must also be independent of *any* rotation of the coordinate system. Then:

$$C_{1111} = C_{1122} + 2C_{1212} \tag{4.4}$$

so the number of independent constants is reduced to two. The two with the most fundamental meanings are the bulk modulus, B, which relates volume changes to pressures, and the shear modulus, G, which relates shape changes to shear stresses at constant volume (for small strains).

4.1 Cubic crystals

In the cubic symmetry system (three constants), consider a volume change brought about by three equal compressive strains applied along the three orthogonal axes:

$$-\epsilon_{11} = -\epsilon_{22} = -\epsilon_{33}$$

Let l be a linear dimension so the volume of a cube is $V = l^3$, and $dV/dl = 3l^2$. The last expression is divided by V, yielding $dV/V = 3(dl/l) = -\epsilon_{11}$. From Equation (4.1), the resulting stress is:

$$\sigma_{11} = (C_{1111} + 2C_{1122})\epsilon_{11} = -(C_{1111} + 2C_{1122})(dV/3V) = -B(dV/V)$$

so, the bulk modulus B is given by:

$$B = (C_{1111} + 2C_{1122})/3 \tag{4.5}$$

For shear strains two moduli are needed. It is convenient to associate one of them with strains parallel to the x_1-axis on the plane perpendicular to the x_3-axis (using Miller indices, in the [100] direction, parallel to the (001) plane). The shear coefficient in this case is C_{1212}. The second one can then be taken for an orientation rotated 45° with respect to the first one, that is, in a direction of the type $\langle 10\bar{1} \rangle$, on a plane of the type $\{101\}$. The coefficient for this is $(C_{1111} - C_{1122})/2 = C^*$.

If a cubic crystal is isotropic, these two coefficients must be equal, so their ratio can be used as a measure of *anisotropy* as proposed by Zener (1948). By convention this anisotropy coefficient is called A and is given by:

$$A = 2C_{1212}/(C_{1111} - C_{1122}) \tag{4.6}$$

When $A = 1$, this is the same as Equation (4.4).

Another shear coefficient that is useful for cubic crystals is that for shears parallel to the $\{111\}$ type planes. These planes have trigonal symmetry (a three-fold rotation axis lies perpendicular to them) so the shear stiffness is the same for any direction that is parallel to them. The coefficient is given by:

$$[3C_{1212}(C_{1111} - C_{1122})]/[4C_{1212} + (C_{1111} - C_{1122})] \tag{4.7}$$

4.2 Strain energy

An elastic strain has an internal energy associated with it. This is predominantly potential energy generated by the work that is done on the solid by the applied tractions in creating the strain. It is stored in the strained structure of the solid, and is recoverable as external work if the tractions are slowly relaxed. A smaller amount, that goes into changing the vibrational frequencies of the atoms in the solid (its temperature), may be recovered, or may be lost as heat, depending on whether the conditions are isothermal or adiabatic (slow or fast).

Storage of energy is an important difference between elastic and inelastic strains. In the case of inelastic strains the shape of the solid changes, but no energy is stored internally.

If the elasticity tensor of a material is known, and the material is elastically strained, a general expression for the strain energy in it can be written. The strain energy density (energy/volume) is equal to the reversible work that has been done to produce the strain. At each point, this energy density is given by:

$$u = \frac{1}{2} \sum_{ijkl} C_{ijkl} \epsilon_{ij} \epsilon_{kl} \tag{4.8}$$

In the isotropic case, for axial tension where $i = j = k = l$, this reduces to the elementary expression: $u = Y\epsilon^2/2$ (Y is Young's modulus).

The local strain energy density can be integrated over the entire volume to obtain the total stored energy:

$$U = \frac{1}{2} \int \left(\sum_{ijkl} C_{ijkl} \epsilon_{ij} \epsilon_{kl} \right) dV \tag{4.9}$$

This integral is of considerable importance in elasticity theory because it is minimized for a set of boundary tractions and/or internal stresses for which the equilibrium conditions are satisfied, as well as the compatibility conditions on the displacements. The displacement field for which this obtains is therefore unique. In other words, if a solution of the differential equations of elasticity meets these conditions, it is unique. There are no other solutions.

4.3 Contracted notation

In most of the literature on the elastic constants, for high symmetry systems, the four subscripts of the tensor notation are contracted to two for convenience. Because of the symmetry of the tensor, this does not lead to ambiguity, and it is more concise. The changes in notation for the cubic case are as follows:

$$
\begin{aligned}
C_{1111} &\rightarrow C_{11} \\
C_{1122} &\rightarrow C_{12} \\
C_{1212} &\rightarrow C_{44}
\end{aligned}
\tag{4.10}
$$

The disadvantage is that the C_{ij} are elements of a matrix, so they do not transform into different coordinate systems like the elements of a tensor. If such transformations are necessary, one must start by going back into the tensor notation.

4.4 Young's modulus

The most common method for determining the mechanical properties of materials is the simple tension test. The elastic response in this test is not ideal, however, because it is a mixture of dilatational and shear strains. Therefore, it is difficult to use simple tension for measuring individual elastic constants.

If the specimen axis is chosen to lie parallel to the $\mathbf{x}_1$-axis, then a simple tension test involves applying a stress σ_{11} parallel to the specimen axis. The specimen responds by elongating longitudinally (ϵ_{11}), and contracting transversely ($-\epsilon_{22}$ and $-\epsilon_{33}$). Since a stress is the applied parameter, the compliance tensor describes the response. Thus:

$$\epsilon_{11} = S_{1111}\sigma_{11} = S_{11}\sigma_{11}$$
$$\epsilon_{12} = S_{1212}\sigma_{11} = S_{12}\sigma_{11}$$
$$\epsilon_{13} = S_{1212}\sigma_{11} = S_{12}\sigma_{11}$$

and $1/S_{1111} = \sigma_{11}/\epsilon_{11} = Y$ (Young's modulus). Similarly, $1/[2(S_{1111} - S_{1212})] = G$ (shear modulus), and $S_{1212} = -\nu/Y$ (ν is Poisson's ratio).

For a material with less than isotropic symmetry, S_{1111} and therefore Y depend on the direction of the tension axis relative to the axes of the symmetry system. To find the variation, direction cosines are used to define the new coordinate system ($\mathbf{x}_1'$, $\mathbf{x}_2'$, $\mathbf{x}_3'$), and the tension compliance is transformed into S_{1111}'. The most common cases of interest are the cubic and the hexagonal systems. In the cubic system, where the direction cosines of a unit vector parallel to the tension axis are c_i, the expression for the variation is:

$$S_{1111}' = S_{1111} - 2[S_{1111} - S_{1122} - 2S_{2323}]\{(c_1c_2)^2 + (c_2c_3)^2 + (c_3c_1)^2\}$$

or, in contracted notation:

$$S_{11}' = S_{11} - 2[S_{11} - S_{12} - 2S_{44}]\{(c_1c_2)^2 + (c_2c_3)^2 + (c_3c_1)^2\} \tag{4.11}$$

For the principal crystallographic directions, the direction cosines (c_1, c_2, c_3) are as follows:

$$
\begin{aligned}
[100] &\quad c_i = 1, 0, 0 \\
[110] &\quad c_i = 1/\sqrt{2}, 1/\sqrt{2}, 0 \\
[111] &\quad c_i = 1/\sqrt{3}, 1/\sqrt{3}, 1/\sqrt{3}
\end{aligned}
$$

and Young's moduli are:

$$
\begin{aligned}
1/Y_{100} &= S_{11} \\
1/Y_{110} &= S_{11} - (1/2)[(S_{11} - S_{12}) - 2S_{44}] \\
1/Y_{111} &= S_{11} - (2/3)[(S_{11} - S_{12}) - 2S_{44}]
\end{aligned}
\tag{4.12}
$$

The condition for isotropy is $Y_{100} = Y_{110} - Y_{111}$. This requires that the term in square brackets be zero, or that the ratio $(S_{11} - S_{12})/2S_{44}$ be unity, and the anisotropy factor may be written (equivalent to Equation (4.6)):

$$A = (S_{11} - S_{12})/2S_{44} \tag{4.13}$$

In a similar fashion, the variation of the shear modulus can be found, since $G(c_i) =$ shear modulus $= 1/S'_{44}$, so that:

$$1/G = S_{44} + 2[2(S_{11} - S_{12}) - S_{44}]\{(c_1c_2)^2 + (c_2c_3)^2 + (c_3c_1)^2\} \tag{4.14}$$

Using the compliance/stiffness inversion formulae:

$$
\begin{aligned}
C_{11} &= (S_{11} + S_{12})/[(S_{11} - S_{12})(S_{11} + 2S_{12})] \\
C_{12} &= -S_{12}/[(S_{11} - S_{12})(S_{11} + 2S_{12})] \\
C_{44} &= 1/S_{44}
\end{aligned}
\tag{4.15}
$$

Or, if the C_{ij} are known, to obtain the S_{ij}:

$$
\begin{aligned}
S_{11} &= (C_{11} + C_{12})/(C_{11} - C_{12})(C_{11} + 2C_{12}) \\
S_{12} &= -C_{12}/(C_{11} - C_{12})(C_{11} + 2C_{12}) \\
S_{44} &= 1/C_{44}
\end{aligned}
\tag{4.16}
$$

For the shear moduli, as was mentioned earlier, two directions must be specified: one for the direction of shear and one for the shear plane. In the cubic system, because of their rotational symmetries, the shear modulus is independent of the shear direction for the {100} and {111} planes. However, this is not true for the {110} plane which has less symmetry. Thus, the shear moduli in the cubic system are given by:

$$
\begin{aligned}
G_{\{100\}\langle *** \rangle} &= C_{44} = 1/S_{44} \\
G_{\{111\}\langle *** \rangle} &= 3C_{44}(C_{11} - C_{12})/(4C_{44} + C_{11} - C_{12}) \\
G_{\{110\}\langle 110 \rangle} &= (C_{11} - C_{12})/2 \\
G_{\{110\}\langle 001 \rangle} &= C_{44}
\end{aligned}
\tag{4.17}
$$

It should also be noted that for tension along a $\langle 110 \rangle$ direction in the cubic system there can be two values of Poisson's ratio. These correspond to the two directions perpendicular to the two mirror planes. One of them can be negative. However, the cross-sectional area still decreases when tension is applied along the two-fold axis. For the three-fold and four-fold axes there is only one Poisson's ratio.

For now, two examples will serve to illustrate anisotropy: copper and diamond. Further discussion of anisotropy will occur later. The stiffnesses of copper (C_{11}, C_{12}, and C_{44}) are 1.68, 1.21, and 0.75 Mbar, respectively (where 1 Mbar (megabar) is equal to 100 GPa). Therefore, its two principal shear moduli are:

$$(C_{11} - C_{12})/2 = 0.235 \text{ Mbar}$$
$$C_{44} = 0.75 \text{ Mbar}$$

The average of these is $\langle G \rangle = 0.49$ Mbar, while the anisotropy coefficient is $A = 3.2$. Furthermore, the bulk modulus of copper is $B = 1.37$ Mbar, so the ratio of the shear to the bulk modulus is $\langle G \rangle / B = 0.36$. Notice also the ratio $C_{44}/C_{12} = 0.62$. If the atoms within copper were bound together by simple bonds between their centers, these ratios would be equal to 1.0, 0.6, and 1.0, respectively. Given the large differences between ideal and actual values, it is clear that the bonding within copper is not simple.

The elasticity ratios for diamond are quite different from those of copper. The stiffnesses are 10.8, 1.25, and 5.76 Mbar, respectively, much larger than for copper. However, diamond is not very anisotropic with $A = 1.2$. Its average shear modulus is $\langle G \rangle = 5.27$ Mbar, while its bulk modulus is $B = 4.43$ Mbar, so the $\langle G \rangle / B$ ratio is 1.19. The ratio $C_{44}/C_{12} = 4.6$ which is quite far from unity (and 7.4 times the value for copper). These ratios are consistent with the idea that there are distinct tetrahedral bonds between the carbon atoms (their symmetry tends to make A equal unity), and that these bonds strongly resist "bending" (it is very rare among materials for G to exceed B). These points will be discussed in more detail in later chapters.

4.5 Cauchy's relations

During the early development of the atomic theory of the elastic constants (Love, 1944), the idea that solids consist of small particles with forces acting along lines connecting their centers was a persistent theme beginning at the time of Newton (ca. 1717), and revived by Boscovich (ca. 1743). An attempt to use it to understand the bending of beams was made by Poisson (ca. 1812), but the first comprehensive advance came from Navier (ca. 1821). He assumed that the forces between the particles are a function only of the radial distance between them, and are independent of the direction of the connecting line (isotropy). In this way he was able to express elastic behavior in terms of just one elastic constant. In other words shear and dilatation are interconnected in Navier's theory (Timoshenko, 1983).

Cauchy made the next step by first introducing the concept of internal *stress* (note that stress is a constructed, rather than a natural variable) through the use of a tetrahedral "free-body" (ca. 1822). This clarified the size of the elasticity tensor (6×6 with 21 independent components), and led to recognition that Navier's molecular theory implied that there should be six equal pairs of these, if each molecule is located at a center of symmetry, leaving 15 independent components for trigonal crystals, two for cubic crystals, and one for isotropic materials. The additional equal pairs are known as the *Cauchy relations*. In the cubic case, the pair is $C_{12} = C_{44}$; for isotropy, $2C_{44} = C_{11} - C_{12}$, so Poisson's ratio is $C_{12}/(C_{11} + C_{12}) = 1/4$. This last conclusion was experimentally tested by Werthheim (ca. 1848). He found that Poisson's ratio was more often 1/3 than 1/4 for the large set of materials he studied. Later Voigt (ca. 1887) studied a variety of crystals loaded in flexure and torsion, and found that only 20% of them satisfied the Cauchy relations even approximately. So it took some 65 years to disprove Navier's theory conclusively. In the light of the quantum chemistry of atomic bonding it would not be expected that pair potentials would describe bonding accurately.

When tractions are applied to a solid they change its shape through shear, and/or its volume through dilatation. If the solid is "elastic", and its temperature is low, the external work that is done in distorting it can be recovered by slowly reducing the tractions. Therefore, the internal energy (strain energy) of the distorted solid is increased by an amount equal to the work done on it. During the distortion process, the magnitude of the traction(s) increases linearly with strain from zero up to the final value (Hooke's Law). Thus the average is half the final stress, and the work done per unit volume (strain energy density) is this times the displacement gradient (displacement per unit length is equal to strain). The reason for the linearity can be understood by considering the behavior near the zero strain, or equilibrium, state.

Suppose a slender cylindrical rod is put into torsion by twisting it through a small angle θ around its axis which is also a symmetry axis of the material of the rod. Then, since θ is the angular displacement per unit length, it is the strain; and the strain energy density $w(\theta)$ is proportional to its square. The Taylor expansion of $w(\theta)$ is formed for the untwisted state, $\theta = 0$:

$$w(\theta) = w(0) + w'(0)\theta + w''(0)\theta^2/2 + w'''(0)\theta^3/6 + \cdots$$

and the twisting torque $\tau(\theta)$, the derivative of $w(\theta)$ with respect to θ, is:

$$\tau(\theta) = w'(0) + w''(0)\theta + w'''(0)\theta^2/2 + \cdots \tag{4.18}$$

where $w'(0) = 0$, since the rod is at equilibrium there, and $w''(0)\theta$ is large compared with $w'''(0)\theta^2/2$ at sufficiently small values of θ. Therefore, for any function, $w(\theta)$, the torque is proportional to θ for small values of θ, and $w''(0)$ is the elastic stiffness coefficient.

In order to see more explicitly how the Cauchy relations arise, the behavior of a small cluster of particles bound by central forces will be analyzed (Feynman, Leighton, and Sands, 1964). The cluster is illustrated by Figure 4.1. It consists of 27 particles in a simple cubic array. Nine of them in the central plane of the cube are shown in the figure. Displacements of the particles from their initial equilibrium positions are small, and the resulting restoring forces depend on the displacements linearly (Hookean). Hence, the interactions can be represented by springs with constants k_i. The interactions of the nearest-neighbor particles have the spring constant k_1, while those of the second-nearest neighbors have the constant k_2. Note that the cluster would have no resistance to shear if there were no k_2. Also, note that only central forces are exerted by k_1 and k_2. The analysis is greatly simplified for a small cluster, but the result has general validity.

We apply a plane strain to the cluster, so $\epsilon_{ij} = 0$, if i, $j = 3$, and the only components of the strain tensor to be considered are ϵ_{11}, ϵ_{22}, and ϵ_{12}. Interactions in the $\mathbf{x}_3$-direction are considered, but no strains. A typical strain-energy term corresponding to an axial strain between particles -1 and -2 (Figure 4.1) which have a separation distance d is $k_1(\epsilon_{11}d)^2/2$. Here the displacement in the $\mathbf{x}_1$-direction is $\epsilon_{11}d = u_1 - \epsilon_{12}d \approx u_1$, since ϵ_{12} is small compared with ϵ_{11}. There are three other similar nearest-neighbor interactions in the $-\mathbf{x}_1$, $\mathbf{x}_2$, and $-\mathbf{x}_2$ directions. The interactions of particle -1 with its nearest neighbors in the $\mathbf{x}_3$

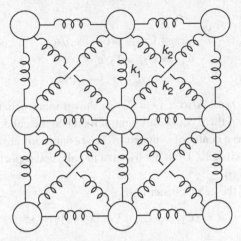

Figure 4.1 The median plane of a model cubic crystal. There is a similar array of nine particles directly in front of this one at the spacing d, and one behind this array also at the spacing d. Hence, the model is a cube containing 27 atoms. The interactions between the atoms are represented by two kinds of springs. Those with spring constants k_1 act between nearest neighbors, while those with constants k_2 act between next-nearest neighbors. The responses are linear for small displacements.

and $-\mathbf{x}_3$ directions do not contribute first-order terms to the strain energy or displacements in the $\mathbf{x}_1$, $\mathbf{x}_2$ plane.

The displacements of the second-nearest neighbors, all 12 of them, contribute to the strain-energy density. The displacements, u_1 and u_2, of particle -3 cause a displacement along the diagonal between particles -1 and -3 equal to the sum of their components in the diagonal direction: $(u_1 + u_2)/\sqrt{2}$. The energy associated with this displacement is:

$$(k_2/2)[(u_1 + u_2)/\sqrt{2}]^2 = [(k_2 d^2)/4](\epsilon_{11} + \epsilon_{12} + \epsilon_{12} + \epsilon_{22})^2$$

and there are three more terms of this type for the particles in the plane of the figure. In addition, in each plane at $\pm d$ in the $\mathbf{x}_3$-direction there are four second-nearest neighbor particles on the edges of the cube. When the displacements of these are resolved into components, and the strain energies are summed the result is:

$$k_2[(\epsilon_{11} d)^2 + (\epsilon_{22} d)^2]$$

The total strain energy, $W(\epsilon)$, for the strained cube, is obtained by adding up these various contributions. There is one term for each pair of atoms, and there is one atom for each d^3 of volume, so the total strain-energy density is:

$$w(\epsilon) = W(\epsilon)/(2d^3)$$

From this, after collecting the terms of the same type, the elastic stiffnesses can be identified.

They are:

$$C_{11} = C_{22} = (k_1 + 2k_2)/d$$
$$C_{12} = C_{21} = k_2/d \qquad (4.19)$$
$$C_{44} = k_2/d$$

All others are equal to zero. Also, $C_{12} = C_{44}$ as shown more generally by Cauchy. This result is a consequence of there being only central forces in the cube, and from the fact that a simple, cubic array has a center of symmetry. These conditions are satisfied quite well by alkali halide crystals such as KCl, but not by most metals, and especially not by covalently bonded crystals.

The bulk modulus of the cubic cluster is:

$$B = (C_{11} + 2C_{12})/3 = (k_1 + 4k_2)/3d \qquad (4.20)$$

and the anisotropy ratio:

$$A = 2C_{44}/(C_{11} - C_{12}) = 2k_2/(k_1 + k_2) \qquad (4.21)$$

Thus, for isotropy (i.e., $A = 1$), $k_1 = k_2$, and the ratio of the shear modulus to the bulk modulus is:

$$C_{44}/B = 3/5 \qquad (4.22)$$

This will be compared with experimental data in a later chapter.

For a more formal approach to the elastic constants, one of the better introductions is in Feynman's treatise (1964). A particularly careful development is in Sines (1969), see also Zener (1948). A convenient source of numerical values is Kittel (1996).

References

Feynman, R.P., Leighton, R.B., and Sands, M. (1964). *The Feynman Lectures on Physics*, Volume II, Chapter 38. Reading, MA: Addison-Wesley.

Kittel, C. (1996). *Introduction to Solid State Physics*, 7th edn. New York: Wiley.

Love, A.E.H. (1944). *A Treatise on the Mathematical Theory of Elasticity*, 4th edn. New York: Dover Publications.

Nye, J.F. (1957). *Physical Properties of Crystals*. Oxford: Oxford University Press.

Sines, G. (1969). *Elasticity and Strength*. Boston, MA: Allyn and Bacon.

Timoshenko, S.P. (1983). *History of the Strength of Materials*. New York: McGraw-Hill, 1953. Republished by Dover Publications, New York.

Zener, C. (1948). *Elasticity and Anelasticity of Metals*. Chicago, IL: University of Chicago Press.

Section III
Elements of electron mechanics

5
Properties of electrons

About 250 years ago, through his mastery of the technology of Leyden jars, Benjamin Franklin, and his research group, discovered one of the most important characteristics of electricity; namely, the charge associated with it is conserved. He reached this conclusion in 1749 when he found that whenever a quantity of charge is taken from a Leyden jar, the charge stored in the jar is reduced by an equal quantity, so the total charge remains constant. He thought of electricity as a continuous fluid of charged particles able to pass through some liquids (e.g., water) and some solids (e.g., metals), but not others (e.g., glass). He identified positive and negative states of charge, one associated with an excess of the electrical fluid, and the other with a deficit of it. His studies were qualitative because there were no adequate instruments for making measurements, but the ingenuity of the studies established his genius. He knew that the force between equal charges declines rapidly with the distance between them, that the resistances of wires increase with the length and inversely with the cross-sectional area, also that electricity passing through a resistive solid generates heat (Cohen, 1941).

Regarding the "electric fluid" contained in glass, Franklin commented (Schonland, 1956), "It seems as if it were of its very substance and essence. Perhaps if that due quantity of electrical fire so obstinately retained by glass could be separated from it, it would no longer be glass. Experiments may possibly be invented hereafter to discover this." And indeed they were, slowly at first, but with acceleration.

Verification of Franklin's hypothesis came only after a century of advances in technology. Methods for conveniently generating, storing, and handling electricity had to be developed as well as methods for evacuating containers. These led eventually to the discovery that hot metal filaments emit electricity which can flow to collectors through vacua (the Edison effect).

The relationships between the various parameters studied by Franklin were made quantitative by Charles Augustin Coulomb and James Prescott Joule. By means of a torsion balance, in 1784, Coulomb showed that two charged bodies attract (or repulse) each other with a force that varies as the inverse square of the distance between them. Later, when methods for measuring quantities of charge became available, it was shown that the force of attraction (or repulsion) also varies as the product of the amounts of charge on the two bodies.

In about 1819, J.J. Berzelius made the first systematic attempt to connect Coulomb's observations with the cohesion of atoms to account for the existence of chemical compounds, and their chemical reactions (Pullman, 1998). He postulated that electropositive, as well as electronegative, atoms exist which can attract one another.

During the middle of the nineteenth century, while he was establishing that mechanical work and heat are equivalent, Joule found that (for linear conductors) the rate at which a conductor becomes heated is proportional to the square of the current that passes through it. The proportionality coefficient is called the resistance.

About 90 years passed before Michael Faraday showed that quantities of electricity could be measured by means of electrochemistry. He found, for example, that if an electric current is passed between two inert electrodes that are immersed in an $AgSO_4$ solution, then silver will deposit on the cathode, and oxygen bubbles will form at the anode. If a mole of the bubbles is collected (32 g), the amount of silver deposited on the cathode will be four moles (432 g, since four parts of hydrogen are needed to make one part of gaseous oxygen). The amount of electricity that passes through the cell in this situation is called the farad, and it equals 96 500 coulomb.

In 1811, Amedeo Avogadro hypothesized that "equal numbers of molecules are contained in equal volumes of all gases under standard conditions." The number in a standard mole is what is now known as Avogadro's constant, but the existence of molecules was not accepted by most chemists until 1858 (47 years later). Therefore, the amount of charge carried by each of the particles in Franklin's electric fluid was not determined until much later. Since a farad represents a mole of electric particles, the charge per particle is given by dividing it by Avogadro's constant. This yields 1.6×10^{-19} coulombs (which approximately equals the current value for the charge of an electron).

New technology was needed before the "particles" of Franklin's electrical fluid could be characterized further. This came in the form of the "Edison effect" which was discovered in 1883, and soon put to scientific use by J.J. Thomson.

During his pursuit of incandescent lighting, Thomas Alva Edison developed the technology for producing high vacua in glass bulbs, and he found that if a hot metallic filament is held in a vacumn tube it emits an electric current that can be collected by a positively charged electrode that resides in the same tube. Thus he isolated the electric "fluid" so it could be studied further. Edison had little personal interest in this effect, presumably because he could not think of a practical use for it, but others soon showed that beams of the "fluid" could be formed, and accelerated by suitable electrodes, and that the beams would cause phosphors like zinc sulfide to luminesce so they could be readily detected. Since the beams originate at hot metallic cathodes, they are called "cathode rays". In modern times they are controlled with great precision in the monitors of computers.

The name "electrons" for the electrical particles was coined in 1891 by G. Johnstone Stoney.

Franklin had already surmised that since the electrical fluid passed through various solids readily, the particles of the fluid must be small, but how small, and how heavy? Are they all the same, or is there a distribution?

A large stride toward answering these questions was made by Sir Joseph John Thomson in 1897. He generated cathode rays by bombarding a cathode with positive ions from a plasma. A beam was formed and accelerated to a velocity, v, by passing it through the voltage drop between two electrodes. The beam was then passed between crossed electric, E, and magnetic, B, fields which imposed oppositely directed forces on it ($-qE$ and qvB where q is the charge, and it is assumed that the beam consists of discrete charged particles). When the fields were adjusted to give zero deflection of the beam, $qE = qvB$, so the velocity could be determined. Then, the electric field was switched off, so under the influence of the magnetic field alone the beam traveled in a circular arc of radius $r = mv/qB$, which could be determined from the deflection of the beam, and the geometry of the apparatus. Then $q/m = v/rB = -1.76 \times 10^{11}$ C/kg.

To determine the values of q and m separately, Faraday's constant plus some assumptions could be used; or the potentials could be reversed to measure q/m for a beam of hydrogen ions since the mass of these was known. However, a direct, and elegant, experiment was devised by Robert Andrews Millikan, and performed in 1909. He arranged for a few tiny droplets of oil to be placed between two horizontal electrodes, and observed them by means of the light they scattered from a beam of light. Then the electric potential between the electrodes was adjusted to equalize the electrical forces, $-qE$, and gravitational forces, mg, on one of the droplets so it was suspended at a fixed vertical position, g being the gravitational constant.

The mass of a droplet depends on its radius, so Millikan used Stokes' Law for the viscous resistance to motion of spheres in liquids to measure the sizes of the droplets. He switched off the potential between the electrodes, and then observed the settling velocities of the droplets until they reached their terminal velocities v_t where the Stokes viscous force just balanced the gravitational force. Under these conditions $r^2 = 9\eta v_t/2g\rho^*$, where ρ^* is the net density between oil and air, and η is the viscosity of air. He observed that the measured values of charge, q', for the droplets were always a small integer times a basic quantity of negative charge. That is: $q' = q$, or $2q$, or $3q$, etc.; the basic quantity, $q = 1.60 \times 10^{-19}$ C is the charge per electron. This result is consistent with Faraday's Law, and has been confirmed by more modern methods.

Combining Millikan's result with Thomson's yields a mass for an electron of 9.11×10^{-28} g. This is 0.000 54 times the mass of a hydrogen atom. Thus the mass is indeed small, consistent with Franklin's supposition.

The nature of electricity and electrons seemed to be understood after Thomson's and Millikan's experiments, but this happy condition did not last long. During 1910–1911, Sir Ernest Rutherford and his students, Hans Geiger and Ernest Marsden, showed that the positive charges in atoms are highly concentrated into nuclei with diameters 10^5 times smaller than an atomic diameter. This posed two questions: (1) what keeps the outer electrons of atoms from combining with the positive nucleus; and (2) what determines the size of an atom? In addition, why do excited atoms (e.g., in plasmas) emit sharply defined spectral lines in well defined groups in which the wavelengths of the individual lines are related to one another by small integral ratios?

Neils Bohr hypothesized an answer to the above questions in 1913. He proposed that the outer electrons of an atom circulate around the positive nucleus in orbits with well defined, constant radii. They stay in these orbits unless a force field acts on them. They can absorb or emit energy only in amounts equal to an integral multiple of a Planck energy quantum, $h\nu$ where h is Planck's constant, and ν is the frequency of the absorbed, or emitted, light. Bohr's theory was quite successful in accounting for the principal observed spectral lines emitted by hydrogen atoms. However, it seemed rather arbitrary, and it did not account for the fine structure, or the hyper-fine structure, that spectroscopists could see in the hydrogen spectrum. Nor did it account for the fact that magnetic fields caused splitting of some of the spectral lines of hydrogen.

The details mentioned above, and many others, began to be resolved when Louis de Broglie suggested in 1923 that, since light has both wave-like and particle-like behavior, perhaps particles with mass, such as electrons, might have the same duality. That is, electrons might have a wave-like behavior with wavelength $\lambda = h/p$ (h is Planck's constant, p is momentum). This relationship was known from the Compton effect to hold for light quanta. de Broglie's hypothesis was soon confirmed in 1927 by the electron diffraction experiments of Davisson and Germer, and it was confirmed for entire atoms in 1929 by Knauer and Stern. So electrons acquired two new properties: a kind of size (wavelength), and a phase (position along a wavelength). These led to rationalization of the sizes and shapes of atoms (and molecules).

de Broglie's idea was a dramatic advance that shook the very foundations of physics as it led to the quantum mechanics of Heisenberg and Schrödinger. Of the idea, Einstein wrote, "I believe it is the first feeble ray of light on this worst of our physics enigmas" (Von Baeyer, 1998). As a result of this postulate, not only the sizes, but also the "shapes", of atoms could be rationalized. The Bohr theory of stationary electronic states for electrons in atoms indicated why atoms have finite sizes, but there was no natural extension of it to account for the shapes of atoms until de Broglie's idea provided it. Since moving waves have phases in addition to amplitudes and wavelengths, electrons confined to circulation around atomic nuclei must have wave functions that close on themselves after each unit of circulation (that is, the initial phase must match the final phase). Thus the circulation is quantized, and since electrons have masses, this means that the angular momenta of the circulating electrons are also quantized. That is, the magnitudes of the angular momenta can only have specific values determined by phase matching. These specific values can be positive or negative depending on the direction of an electron's circulation: clockwise or counterclockwise. Overall, then, de Broglie's idea grew into three quantized properties for the orbitals of atomic electrons, and therefore three quantum numbers to indicate the magnitudes of the properties. The principal quantum number, n, gives the energy and size of an electron orbital. The angular momentum quantum number, l, gives the symmetry (shape) of the orbital. If the atom is immersed in a magnetic field, there is a third, magnetic quantum number, m, but it will not concern us here (Born, 1989).

Soon, Erwin Schrödinger (1925) proposed that the wave-like characters of particles might be described by means of an equation similar to the differential equations that describe acoustic, elastic, and electromagnetic waves. When this was applied to the simplest of atoms, namely the hydrogen atom, it was brilliantly successful in accounting with quantitative precision for all of the spectroscopy, except for the finest details.

However, some difficulties remained. (1) In a multi-electron atom, why do not the electrons seek the lowest energy for the system by all dropping into the spherical orbitals that have the lowest energy? (2) Why do neutral atoms resist penetration by one another?

A new postulate was needed. A simple one by Pauli came to the rescue: two, and only two, electrons can occupy a given quantum state (i.e., a state with a given energy level, angular momentum, and spacial extent). This Pauli Principle, together with Schrödinger's wave equation, accounted extremely well for the patterns of the Periodic Table of the Elements.

Concern remained about why the rule was two electrons in a given orbital, why not one, or three? Also, high resolution spectroscopes showed splitting of some of the optical emission lines that could not be accounted for, particularly in the presence of a magnetic field (the Zeeman effect). These concerns were addressed by Goudsmit and Uhlenbeck by postulating a peculiar new property of electrons, namely "spin", and an associated magnetic moment. Spin can have only one of two values, up or down. With spin, the Pauli Principle (1925) postulated that atomic orbitals that contain paired electrons (net spin of zero) are fully occupied, so any further electrons in the system must go into orbitals of higher energy. As a result, when the simple spherical orbitals become filled with pairs, electrons begin to occupy the more complex, higher energy, orbitals. Without this principle, chemistry would be barren indeed. However, although the wave functions that satisfy the Schrödinger equation have simple shapes (i.e., spherical symmetries and exponential decays) for the lowest energy states, as the kinetic energy of the electrons increases the shapes of the distributions become more complex (for example, six lobes directed along three orthogonal axes). It is this variety of shapes that leads to the richness of the geometries of molecules and crystals. In the present context, this is important because it accounts, for example, for the difference between the hardnesses of pure metals and those of pure semiconductors. It also plays an important role in determining the shear stiffnesses of solids.

By combining de Broglie's idea with Planck's Principle ($E = h\nu = hc/\lambda$) another conceptual leap was made by Heisenberg (1927). If an electron has a wavelength associated with it, then neither its position, nor its velocity, are precisely defined. However, if the standard deviation from its mean position is Δr, and the standard deviation from its mean velocity is Δv, then Heisenberg showed that their product is equal to a fixed quantity (h is Planck's constant, and m is the electron mass):

$$(\Delta r)(\Delta v) = h/4\pi m \qquad\qquad (5.1)$$

Now, as Δr decreases defining the position more accurately, the kinetic energy increases since Δv must increase.

If the electron is in an atom, and therefore is being attracted to a positive nucleus, its electrostatic energy is decreased when its distance from the nucleus decreases, but its kinetic energy increases faster than its electrostatic energy decreases. This causes the atom to resist compression, and allows us to make a simple calculation of the bulk moduli of metals as will be shown shortly.

In 1922, Otto Stern and Werner Gerlach performed a landmark experiment which indicated that electrons have still another property! From a vapor of silver atoms they formed a beam of atoms in a vacuum. The beam was passed through a magnetic field gradient oriented transverse to the direction of motion of the atomic beam. The latter was then condensed onto a cold glass slide. However, instead of one condensation spot, they found two! One was as a result of a small deflection to the left of the center of the beam, and the other from a small deflection to the right. Silver is monovalent, so this behavior became associated with the valence electron having a magnetic moment with two oppositely directed orientations. On the basis of the forces that the field gradient exerted on the beam(s), spin quantum numbers were assigned to the valence electrons that are consistent with the quantum numbers previously determined from optical spectroscopy. These are $+1/2$ and $-1/2$.

For the purposes of this book, this completes the list of characteristics of electrons. There are some others such as the existence of a sister particle, the positron, but they have little effect on strength properties, so they will not be discussed.

Summing up, electrons have the following properties: mass, charge, wavelength (dependent on energy), phase, and magnetic moment (spin). If they are associated with p, d, f, ... orbitals, they also have angular momentum associated with their non-spherical orbitals (and in addition to their self-spins). All of these properties play important roles in determining the strengths of materials. As will be discussed in subsequent chapters, the charge yields the basic electrostatic forces of attraction and repulsion; the wavelength determines kinetic energy, atomic size, and atomic shape; the self-spin affects the distribution of charge between atoms and therefore the bonding, as well as the anti-bonding.

References

Born, M. (1989). *Atomic Physics*, 8th edn., trans. J. Dougall, revised R.J. Blin-Stoyle and J.M. Radcliffe. New York: Dover Publications.

Cohen, I.B. (Editor) (1941). *Benjamin Franklin's Experiments*. Cambridge, MA: Harvard University Press.

Pullman, B. (1998). *The Atom in the History of Human Thought*, trans. A. Reisinger, p. 206. New York: Oxford University Press.

Schonland, B.F.J. (1956). Benjamin Franklin: natural philosopher, *Proc. R. Soc. London, Ser. A*, **235**, 433.

Von Baeyer, H.C. (1998). *APS News*, May, p. 3.

6

Quantum states

The most simple atom is hydrogen consisting of a positive proton at the center, and one negative electron swarming around it. While the proton is concentrated at the center, the electron swarms in a pattern that depends on its total energy, and its angular momentum. In the lowest energy state of the hydrogen atom, the shape of the pattern is simply a sphere, but in higher energy states the pattern resembles a dumbbell, or a three-dimensional cloverleaf. For atoms with more than one electron, each electron has its own pattern, or state.

The fundamental postulate that determines the behavior of atomic electrons is that of de Broglie (1924). This states that an electron of mass, m, and velocity, v, and therefore with momentum, $p = mv$, has a wavelength, λ, associated with it that is given by $\lambda = h/p$, where h is Planck's constant. If $T = mv^2/2$ is the kinetic energy, then the expression for the wavelength can also be written $\lambda = hm/2T$, so the wavelength depends inversely on the kinetic energy. Thus short wavelengths mean high energies and vice versa. Initially, this seemed to be a rather arbitrary proposal, but it has been verified by many, and differing, experiments, and has led to vast numbers of quantitatively precise interpretations of phenomena. The first verification came from the observations of Davisson and Germer (1927) that electrons of known kinetic energy can be diffracted from the surfaces of nickel crystals.

An important aspect of the wave-like behavior is that boundary conditions impose restrictions on what wavelengths are acceptable. Thus, for example, guitar strings can only vibrate at a small set of wavelengths, not at an arbitrary wavelength. The longest acceptable wavelength (lowest frequency) is equal to twice the distance between the two ends of a string. The next longest wavelength is half as long as the lowest one (twice the frequency). Then comes the wavelength that is one-third as long as the lowest one, and so forth. Between these allowed vibrational modes, others are not allowed because the boundary conditions (pinning points at each end) cannot be satisfied. In other words, the vibrational wavelengths, frequencies, and kinetic energies of guitar strings are quantized. Their wavelengths are very much larger than those of electrons in atoms, of course. Forming a bridge between quantum mechanics and wave mechanics was the great contribution of Erwin Schrödinger. He pointed out that since electrons and other particles have wave-like character, they should obey a differential equation similar to those for acoustic and electromagnetic waves. To understand this, some elements of wave mechanics as it applies to other phenomena may be helpful.

6.1 Wave-like fields

In acoustics (in gases, liquids, and solids) the displacement, $u(x)$, of a point for a one-dimensional wave, is described by the wave equation:

$$\frac{\partial^2 u}{\partial x^2} = \frac{1}{v_s^2}\left(\frac{\partial^2 u}{\partial t^2}\right) \tag{6.1}$$

where t is time and v_s is the velocity of sound. The term on the left-hand side describes the curvature, or spacial shape, of the displacement at any particular time. The right-hand side describes how the displacement changes with time at a particular spacial point, the first factor being the velocity factor, and the second the acceleration. The velocity factor, v_s^2, increases with the stiffness of the medium, and decreases with its mass density. The solution of this equation is periodic in both space and time.

In gases and liquids, only longitudinal displacements (compressions and rarefactions) exist so $u = f(x, t)$, a function of x and t only. In solids both longitudinal and transverse (shear) displacements exist so $u = u_x = f(x, t)$, or $u = u_{xy} = g(x, y, t)$ for plane polarized waves, or both. And, in this case there are two values of v_s, one for longitudinal waves and one for shear waves.

For electromagnetic waves, including light, there are also two fields, the electric, E, and the magnetic, B. If F represents either of them, the wave equation is:

$$\frac{\partial^2 F}{\partial x^2} = \frac{1}{c^2}\left(\frac{\partial^2 F}{\partial t^2}\right) \tag{6.2}$$

where c is now the velocity of light, and, according to Einstein, $c^2 = E/m$ where E is the total energy and m is mass. Planck's equation yields $E = h\nu$, so the frequency is $\nu = (mc^2)/h$. Again, the solutions are periodic in time and space. The intensity, I, is proportional to the square of the amplitude, that is, $I \propto F^2$.

For an electron, the field variable is Ψ which is the probability amplitude, analogous with F the electromagnetic field amplitude, and u the displacement amplitude. Then Ψ is a measure of the quantum amplitude field, while Ψ^2 is the probability density. The wave equation for the quantum field (using the de Broglie relation $\lambda = h/p$, so the velocity is E/p) is:

$$\frac{\partial^2 \Psi}{\partial x^2} = \left(\frac{p}{E}\right)^2 \left(\frac{\partial^2 \Psi}{\partial t^2}\right) \tag{6.3}$$

Often, the desired probability amplitude wave is a standing wave which does not depend on x, but changes with time t. Then a solution of Equation (6.3) is:

$$\Psi = \psi e^{-i\omega t} \tag{6.4}$$

where $\omega = 2\pi\nu$, the circular frequency. In this equation, $\psi = \psi(x, y, z)$, is a function of position, but not of time, whereas $\Psi = \Psi(x, y, z, t)$. Substituting Equation (6.4) into Equation (6.3), and remembering that the kinetic energy is $T = p^2/2m$, while $E = h\nu$, yields:

$$\frac{\partial^2 \psi}{\partial x^2} = -\left(\frac{8\pi^2 mT}{h}\right)\psi \tag{6.5}$$

but the total energy is the sum of the kinetic and potential energies, $E = T + V$, so Equation (6.5) can be written:

$$\frac{\partial^2 \psi}{\partial x^2} + \frac{8\pi^2 m}{h}(E - V)\psi = 0 \tag{6.6}$$

which is the standard form of the time-independent Schrödinger wave equation. Note that this equation contains Planck's constant, whereas Equation (6.2) does not.

The intensity of an electrical wave equals the square of the electric field E. This is natural since the field amplitude F can take negative as well as positive values, but the intensity E^2 is always positive (the intensity of light, for example).

Similarly, the probability amplitude Ψ is squared to obtain the probability density, P, at position x. However, since solutions of wave equations are conveniently written in terms of $e^{-i\omega t}$, a complex number, the square of Ψ is often written as the number times its complex conjugate, that is, in terms of the absolute value squared. Thus:

$$\Psi^2 = \Psi\Psi^\star = (\Psi e^{-i\omega t})(\Psi e^{+i\omega t}) = |\Psi|^2$$

In describing the behavior of a given electron, the totality of its probability distribution must equal unity since the electron must be somewhere. Therefore, probability amplitudes must be normalized by integrating Ψ^2 over all positions, and setting the result equal to unity. If the normalization factor is called N, and $\Psi = N\psi$, then, for the one-dimensional case:

$$N^2 \int \psi\psi^\star \, dx = 1 \tag{6.7}$$

Atomic systems sometimes change with time, so a more general time-dependent Schrödinger equation is needed. Using the time derivative of Equation (6.4):

$$\frac{-h}{2\pi i}\left(\frac{\partial \Psi}{\partial t}\right) = h\omega\psi = E\psi \tag{6.8}$$

so that changes of Ψ follow first-order kinetics in which the rate of change of the amplitude is proportional to the current value of the amplitude. Therefore, the integral of the amplitude for simple cases is just an exponential function with an imaginary exponent, and Equation (6.6) becomes:

$$\frac{\partial^2 \Psi}{\partial x^2} + \frac{8\pi^2 m}{h^2}\left[\frac{h}{2\pi i}\frac{\partial}{\partial t} - V(x)\right]\Psi = 0 \tag{6.9}$$

All of these equations can be readily generalized for three dimensions.

The wave equation for the probability amplitude (Equation (6.3)) is similar to that for the electric field (Equation (6.2)), and the acoustic field (Equation (6.1)) with important

exceptions. Neither the wave form of the amplitude nor the amplitude velocity are necessarily constant. Also, the position operator, **x**, and the momentum operator, **q**, do not
commute. That is,

$$\mathbf{pq} \neq \mathbf{qp}$$

Instead:

$$\mathbf{pq} - \mathbf{qp} = h/2\pi i \qquad (6.10)$$

where $\mathbf{q} = x$, and $\mathbf{p} = (h/2\pi i)\partial/\partial x$. Equation (6.10) was Heisenberg's original postulate
with **p** and **q** being matrices in three dimensions. From this deceptively simple relationship,
all of quantum mechanics flows. For macroscopic Newtonian mechanics, Planck's constant
$h = 0$, so **p** and **q** then commute. Remarkably enough, Schrödinger's equation is equivalent
to the commutation condition, Equation (6.10).

In an atom, electrons are confined to a certain amount of volume. Because of Heisenberg's
Principle, this volume is not precisely defined, but its average value is known from atomic
scattering experiments, and from the known lattice parameters of crystals. From symmetry
considerations, atoms are roughly spherical, but also have other shapes including ellipsoids,
spheres with "bumps" at each end of their polar axes, spheres with three bumps on their
equators lying 120° apart, spheres with four bumps evenly distributed at tetrahedral angles
(109.5°) from each other, spheres with six bumps, each at the ends of three orthogonal axes,
spheres with eight bumps in an octahedral array, and so forth. A marvel of the Schrödinger
equation is that its solutions yield all these shapes for a quite simple form of the potential
energy, $V(x)$, that is, for the spherically symmetric Coulomb potential provided by the
atomic nucleus. If the atomic number is Z, its unscreened charge is $-Zq$, where q is the
negative electron charge, and $V(r) = -Zq/r$ with r the radial distance from the nucleus.
How the various shapes come about will be shown here in steps in order to indicate how
the solutions of the equation are affected by boundary conditions (Atkins, 1983). The first
case to be considered is that of a particle whose motion is confined to a ring, that is, to
motion in a plane at a constant radial distance from the center of rotation. This will serve
to introduce the idea of quantized angular momentum which is the basis of the differing
shapes of atoms.

6.2 Particle on a ring

From classical mechanics, a particle of mass m that moves around a center at a distance
r has a moment of inertia $I = mr^2$. For a particle on a ring (Figure 6.1), the motion is
two-dimensional, involving both x and y, so the expression for the kinetic energy is:

$$\frac{h^2}{8\pi m}\left(\frac{\partial^2 \psi}{\partial x^2} + \frac{\partial^2 \psi}{\partial y^2}\right) \qquad (6.11)$$

The potential energy V is zero, since by definition the particle is not allowed to move
off the ring. The configuration has circular symmetry, so it is convenient to use polar

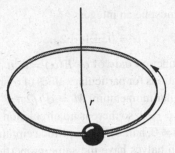

Figure 6.1 Particle on a ring. The particle circulates around an axis at a fixed distance, r.

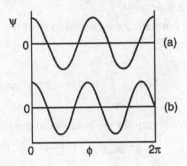

Figure 6.2 Quantization of angular probability amplitudes: (a) satisfactory function, starts an orbit at $\phi = 0$ and finishes it at $\phi = 2\pi$, with the same amplitude; (b) unsatisfactory function, starts and finishes an orbit at different amplitudes.

coordinates (r, ϕ). Then, $x = r \cos \phi$, and $y = r \sin \phi$, where r is the distance of the particle from the center of the ring, and ϕ is the angular position of the particle along the ring. The moment of inertia, rather than the mass alone, determines the kinetic energy in this case, and its value is $I = mr^2$. Thus, in polar coordinates, the kinetic energy becomes:

$$\frac{h^2}{8\pi^2 I} \left(\frac{\partial^2 \psi}{\partial \phi^2} \right) \tag{6.12}$$

and the Schrödinger equation becomes:

$$\frac{\partial^2 \psi}{\partial \phi^2} = - \left(\frac{8\pi^2 I E}{h^2} \right) \psi \tag{6.13}$$

the solution being:

$$\psi(\phi) = A e^{i\alpha\phi} + B e^{-i\alpha\phi} \qquad \alpha = \left(\frac{8\pi^2 I E}{h^2} \right)^{1/2} \tag{6.14}$$

Here α can have any value, but there is another requirement that restricts it to have only special values. This requirement is that $\psi(\phi)$, and its derivatives, must be continuous upon going around the ring. Otherwise there would be places where $\psi(\phi)$ would have more than one value (Figure 6.2). This continuity requirement is satisfied if $\psi(\phi) = \psi(\phi + 2\pi)$; since

$e^{2\pi i \alpha} = 1$, this means that α must be an integer:

$$\alpha = 0, \pm 1, \pm 2, \ldots$$

Then, since the energy is a function only of α, $E(\alpha) = \alpha^2 h^2 / 8\pi^2 I$, and it can have only discrete values. The configurations for particular values of α are called energy states. α also determines the particle's angular momentum, $M = \alpha h / 2\pi$.

For the lowest energy state $\alpha = 0$, so there is no circulation of the particle, and therefore no angular momentum. For $\alpha \neq 0$, half of the amplitude circulates clockwise, $e^{i\alpha\phi}$, and half counterclockwise, $e^{-i\alpha\phi}$. Both halves have the same properties except for the sign. If we write for the clockwise probability amplitude, $\psi_+ = A e^{i\alpha\phi}$, then the probability for the particle to be in this state is $\psi_+ \, \psi_+^* = A^2$.

To find the value of A, the normalization condition (that the total probability is unity) is used (Equation (6.7)):

$$\int_0^{2\pi} \psi \psi^* \, d\phi = \int_0^{2\pi} A^2 \, d\phi = 2\pi A^2 = 1 \qquad (6.15)$$

So $A = 1/\sqrt{(2\pi)}$, and the probability for the position of the particle on the ring is:

$$\psi \psi^* = A^2 = 1/(2\pi)$$

which is independent of ϕ, making the distribution of the position of the particle uniform around the ring. This differs from classical mechanics, of course, where there would be a localized particle circulating around the ring.

6.3 Particle on a sphere

In an atom, the motion of an electron is not restricted to a plane (ring). Since its total energy depends only on its distance from the positive nucleus, it might be concluded that it can roam anywhere on the surface of a sphere of radius r. However, because of quantum conditions its roaming is limited to particular places. This leads to a complex mathematical description which is why the case of planar motion on a ring was considered first. Now, before the full complexity is considered, an intermediate case will be discussed in order to minimize the complications in going from two to three dimensions. The spherical case that will be initially discussed differs somewhat from an atom in that it will be assumed that there is no interaction (potential energy) between the particle on the surface of the sphere and the center of the sphere. The particle is arbitrarily confined to motion on the spherical surface.

To describe motion on a sphere, a new angle θ is needed (Figure 6.3). In the intermediate case, the particle then lies at a fixed distance r from the center of the sphere, and at longitude ϕ, as well as latitude θ. In these coordinates:

$$x = r \sin\theta \cos\phi$$
$$y = r \sin\theta \sin\phi$$
$$z = r \cos\phi$$

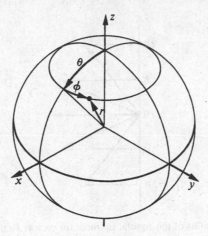

Figure 6.3 Definition of the spherical coordinates: r radius, θ latitude, and ϕ longitude.

and since $r = $ constant, the r derivatives drop out of the Laplacian operator. It becomes $\nabla^2 = (1/r^2)\Lambda^2$, where $\Lambda^2 = f(\theta, \phi)$ is the Lagrangian. Thus the Schrödinger equation becomes:

$$\Lambda^2\Psi(\theta, \phi) = -\left(\frac{8\pi^2 EI}{h^2}\right)\Psi(\theta, \phi) \tag{6.16}$$

The solution of this is separable into two angular parts with $\Psi(\theta, \phi) = \Theta(\theta)\,\Phi(\phi)$ where the equation for Φ is:

$$\frac{d^2\Phi}{d\phi^2} = -m^2\Phi \tag{6.17}$$

which has solutions like those for the particle on a ring (for motion around the polar axis):

$$\Phi = e^{im\Phi} \tag{6.18}$$

Notice that the symbol, α, that was previously used for the angular momentum is now m. The m stands for "magnetic angular momentum" for historical reasons. The "mechanical angular momentum" quantum number will be called l ("el"), shortly.

The additional coordinate, θ, leads to further quantization consisting of a series of ring-like states at discrete values of the latitude θ (Figure 6.4). The equation for $\Theta(\theta)$ is more complicated than the one for $\Phi(\phi)$, and will not be discussed in detail here. Interested readers are referred to other books such as Pauling and Wilson (1983).

Equation (6.16) has the same form as the standard differential equation for functions known as spherical harmonics, $Y_{lm}(\theta, \phi)$. The equation that these functions satisfy is:

$$\Lambda^2 Y_{lm}(\theta, \phi) = -l(l + 1)Y_{lm}(\theta, \phi) \tag{6.19}$$

and Y_{lm} is separable: $Y_{lm}(\theta, \phi) = \Theta_{lm}(\theta)\Phi_m(\phi)$ with the $\Phi_m(\phi)$ being given by Equation (6.17). The $\Theta_{lm}(\theta)$ are called associated Legendre functions. For $l = 1, 2, 3$ and $m = 0$,

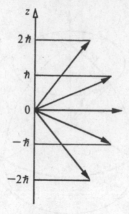

Figure 6.4 Permitted orientations of the angular momentum vector, $\mathbf{J}(l)$, relative to a reference field vector, $\mathbf{z}$. The values of l, and therefore $\mathbf{J}$, are fixed. In this case, $l = 2$, so $m = 0, \pm 1, \pm 2$ is the magnetic quantum number. The component along $\mathbf{z}$ interacts with the field.

the first few are given by:

$$\Theta_{00} = (1/2\pi)^{1/2} \qquad \text{circular shape}$$
$$\Theta_{10} = (3/4\pi)^{1/2}\cos\theta \qquad \text{figure of eight shape}$$
$$\Theta_{20} = (5/16\pi)^{1/2}(3\cos^2\theta - 1) \qquad \text{cloverleaf shape}$$

From the Schrödinger equation then, various shapes emerge for the spacial distributions of the probability amplitudes without further assumptions. Also, the energy and the angular momentum emerge. Comparison of Equations (6.16) and (6.19) shows that the energy is:

$$E = \frac{h^2}{8\pi^2 I} [l(l + 1)] \qquad (6.20)$$

where $l = 0, 1, 2, \ldots, (n - 1)$, and the angular momentum is $J = I\omega$, with ω the rotational frequency, $\omega = 2\pi\nu$. Therefore, the rotational energy is $E_{\text{rot}} = J^2/2I$. Substituting into Equation (6.20) gives an expression for the angular momentum:

$$J = \frac{h}{2\pi} [l(l + 1)]^{1/2} \qquad (6.21)$$

So l is the angular momentum quantum number, and it can have the values:

$$l = 0, 1, 2, \ldots, n$$

For a total amount of energy determined by the value of n, if $l = 0$, there is no angular momentum, so $\Theta_{lm}(\theta) = \Theta_{00} = $ constant and the symmetry of the probability amplitude is spherical. However, for example, if $n = 2$, so $l = 2$ is permitted, then the angular momentum, according to Equation (6.21) is $\mathbf{J} = (\sqrt{6})\hbar$. This is a vectorial quantity as in classical mechanics, but it is unusual in that it can have only particular directions relative to a reference vector in space. The reference might be a magnetic field, $\mathbf{H}$, or an electric field, $\mathbf{E}$. Historically, the magnetic case was discovered first, so it is called the magnetic

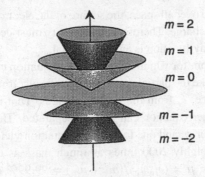

Figure 6.5 Cones on which the angular momentum vector may lie.

quantum number, and is designated m. The particular values of θ it can have are $m = 0$, $\pm 1, \pm 2, \ldots, \pm l$. This is illustrated by Figure 6.4 which shows how the orientation varies with m, at constant l. As the orientation of the angular momentum vector varies, its component along the reference direction varies. This means, of course, that the interaction with the reference field vector varies, and hence the interaction energy. For each value of m, although the total angular momentum l is constant, the permitted components in a particular direction are determined by m.

If there is no magnetic, or electric, field present to affect the angular momentum, the spacial quantum number m is zero.

For a given set of n, l, and m in the presence of a reference field, the value of θ is fixed, but ϕ can have any value, so the angular momentum vectors lie on cones centered on the reference vector (Figure 6.5).

The second derivatives (with respect to x, y, z, or r, θ, ϕ) of the probability amplitudes represent curvatures. Thus the kinetic energy is proportional to the curvature of the probability amplitude. This is consistent with the highest frequency Fourier component of the amplitude (shortest wavelength), and follows from Planck's Principle, $E = h\nu$ which gives high energies for short wavelengths.

The total energy of a state depends only on the value of the principal quantum number n, not on l or m.

6.4 The most simple atom (hydrogen)

With the particle on a ring, and the particle on a sphere as background, atoms are next. The simplest case is that of hydrogen (one nuclear proton and one orbital electron). Note that the word orbital is used instead of "orbiting" because quantum (wave) mechanics does not consider the electron to be encircling the proton like a planet encircling the sun. Instead, the electron is considered to exist only in states of definite energy that are given by solutions of Schrödinger's equation, and which change relatively slowly under the influence of various force fields. The probability density of an electron provides a distribution function that shows how the electron density is distributed over the volume of an atom, and what the shape of the volume is. When integrated over all space, the square of the electron's probability amplitude must equal one electronic charge. Therefore, the density must decay rapidly beyond some perimeter. Usually the decay function is an exponential.

Hydrogen is the only atom for which Schrödinger's equation can be solved analytically. In the hydrogen atom (unlike the particle on a sphere), the electron is no longer constrained to lie at a constant distance, r, from the proton nucleus. That is, r becomes a variable. Fortunately, the Schrödinger equation can still be separated. The separation yields three equations: one for r, one for ϕ, and one for θ. Each equation yields one quantum number.

Since the proton has roughly 2000 times as much mass as the electron, the reduced mass for the two particles, $\mu = (m_e m_p)/(m_e + m_p) = 99.99\%$ of the electron mass. So the electron mass can be used in the equations (rather than the reduced mass) without introducing much error.

The potential energy is the Coulombic energy of the electron–proton pair; that is, $V(r) = -q^2/r$ (cgs or atomic units), and the Schrödinger equation in spherical polar co-ordinates becomes:

$$\frac{1}{r}\left(\frac{d^2}{dr^2}\right)(r\psi) + \frac{1}{r^2}\Lambda^2\psi + \left(\frac{4\pi^2 mq^2 r}{h^2}\right)\psi = -\left(\frac{8\pi^2 mE}{h^2}\right)\psi \qquad (6.22)$$

This can be separated into a radial part with the probability amplitude $R(r)$, and an angular part with the probability amplitude $Y(\phi, \theta)$. Then: $\psi(r, \phi, \theta) = R(r)Y(\phi, \theta)$.

Using Equation (6.19), $\Lambda^2 Y = -l(l+1)Y$ ($l = 0, 1, 2, \ldots, n$), and substituting into Equation (6.22) with cancellation of the Y, yields an equation for the radial probability amplitude:

$$\frac{1}{r}\left(\frac{d^2}{dr^2}\right)(rR) + \left[\frac{4\pi^2 mq^2 r}{h^2} - \frac{l(l+1)}{r^2}\right]R = -\left(\frac{8\pi^2 mE}{h^2}\right)R \qquad (6.23)$$

In this equation, unlike the corresponding one for the "particle on a sphere", the r-derivative appears because the electron is not constrained to lie at a constant distance from the proton. Making the substitution, $\Omega = rR$, this equation becomes:

$$\frac{d^2\Omega}{dr^2} + \frac{8\pi^2 m}{h^2}\left[\frac{q^2}{r} - l(l+1)\frac{h^2}{8\pi^2 mr^2}\right]\Omega = -\left(\frac{8\pi^2 E}{h^2}\right)\Omega \qquad (6.24)$$

where the term in square brackets is an effective potential with two parts. The first part is an attractive Coulombic potential, and the second part is a repulsive angular momentum potential (analogous with the centrifugal energy of a ball on a string being swung in a circle). Thus the forces on the electron are balanced with the electrostatic attraction of the nuclear proton pulling inward, and the angular acceleration of the electron pushing outward, analogous with the Bohr theory.

For the lowest energy state, $l = 0$, and there is no angular momentum (unlike the classical case of electrons orbiting a center of positive charge). This is called the 1s state. There are also excited states with no angular momentum, labelled 2s, 3s, 4s, etc. All of these are spherically symmetric (the "s" is a useful mnemonic, but comes from early spectroscopy,

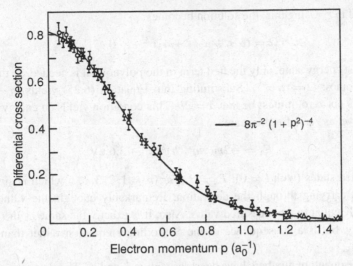

Figure 6.6 Experimental electron density distribution, determined by measuring the momenta of both incident and scattered electrons for individual scattering events from hydrogen. Symbols correspond to three incident energy levels: × 1200 eV, ○ 800 eV, △ 400 eV. The solid line corresponds to the theory of the differential cross-sections for the 1s state of hydrogen. Data of Vos and McCarthy (1995).

not from wave mechanics). The probability amplitudes of the ns states oscillate as their averages drop off exponentially from the central nucleus. Vos and McCarthy (1995) have verified this experimentally for the 1s state ($n = 1, l = 0$) of hydrogen (Figure 6.6). Their energies can be found from further consideration of Equation (6.24). To do this in a complete way is rather cumbersome so a simplified version will be used here. The first step is to make Equation (6.24) more compact by collecting the constants into master constants, and by considering only the lowest energy state (the ground state) so $l = 0$. Then the equation becomes:

$$\frac{\partial^2 \Omega}{\partial r^2} - \left(b^2 - \frac{a}{r}\right)\Omega = 0 \tag{6.25}$$

where $a = 8m(\pi q/h)^2$, and $b^2 = 8mE(\pi/h)^2$. At large r, the second term in parentheses drops out and the equation becomes:

$$\frac{\partial^2 \Omega}{\partial r^2} - b^2\Omega = 0 \tag{6.26}$$

This is similar to Equation (6.13), and the solution is similar:

$$\Omega = e^{-br} \tag{6.27}$$

where the minus sign has been chosen in order to allow normalization. Going back to Equation (6.25), Equation (6.27) needs a function of r in it in order to satisfy the part that describes the Coulomb potential (a/r). The simplest general function that might be used is a polynomial in r. Using this, the solution becomes:

$$\Omega = (\alpha_1 r + \alpha_2 r^2 + \alpha_3 r^3 + \cdots) \, e^{-br} \tag{6.28}$$

For the lowest energy state, only the first term in the polynomial is needed, so the proposed solution becomes: $\Omega = \alpha r e^{-br}$. Substituting into Equation (6.25) yields $(a - 2b)r = 0$, and since r is not zero, it must be that $a = 2b$. This condition yields an energy (taking the ionization level as zero) for this state:

$$E_1 = -2m(\pi q^2/h)^2 = -13.6 \text{ eV} \tag{6.29}$$

For the excited states (with $l = 0$): $E_n = E_1/n^2$ ($n = 1, 2, 3, \ldots$) which is simply stated here without carrying through the derivation. Remarkably enough, the value for E_1 is almost exactly equal to the measured value. Also, it is exactly the same as the value from Bohr's theory, but is a consequence of the Schrödinger equation rather than arbitrarily setting $n = 1$.

So far three quantum numbers have been derived: n, l, and m. The first, n, is related to r, and through the constant b to the "size" of the atom. The decay distance, r is given by:

$$r = n^2 a_0 \tag{6.30}$$

where a_0 is the Bohr radius, $a_0 = (1/m)(h/2\pi q)^2 = 0.529 \text{ Å}$, m and q being the mass and charge of the electron, respectively, and h is Planck's constant. Equation (6.30) indicates that the "size" increases rapidly with the principal quantum number, n. Figure 6.7 illustrates this for the first few values of n. The figure also shows how the charge density varies with the same values of n.

The second quantum number, l, quantizes the angle ϕ, changes the "shape" from spherical symmetry, and thereby determines the angular momentum.

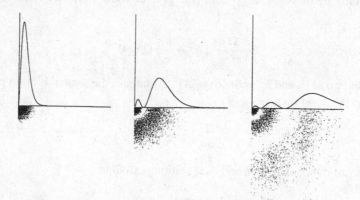

Figure 6.7 Radial distribution functions for the spherically symmetric electron densities of the $n = 1, 2$, and 3 (1s, 2s, and 3s) quantum states. The peaks move outward along the horizontal (r) axis in proportion to n^2. Also shown are dot representations of the electron densities for one quadrant in each case.

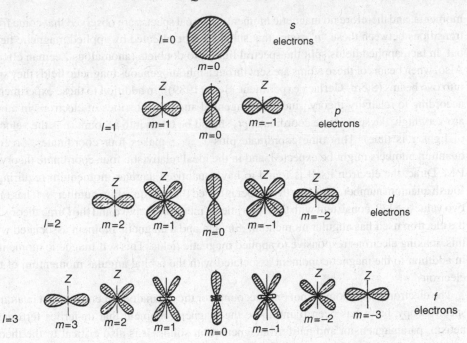

Figure 6.8 Schematic "shapes" of the probability amplitude functions for the first four values of the angular momentum quantum numbers, l, and the corresponding magnetic quantum numbers, m. The correspondence between the spectroscopic term names s, p, d, and f, and the values of $l = 0$, 1, 2, and 3 are also shown.

The third, $\pm m$, determines the "latitude" of the orbital. If the atom is immersed in a magnetic field, this determines the effective magnetic moment of the orbital motion. These effects are illustrated by Figure 6.8 which also indicates the letter names (s, p, d, and f) that were used by spectroscopists to organize the types of spectral lines they observed, long before the advent of quantum numbers. The use of these letter designations has persisted, so the correspondences between them and values of the angular momentum quantum numbers (the l values) should be remembered.

Except for the case of $n = 1$, there is more than one probability amplitude function that has the same energy with different angular momenta. Such energy levels are said to be *degenerate*. Applied **H** or **E** fields split such levels into their components. This causes the spectral lines associated with these levels to split into closely spaced clusters of lines.

There is one more coordinate that plays a role.

6.5 Electron spin

Hydrogen-like atoms such as Li, Na, K, Cs, and Rb have electronically saturated ion cores plus one outermost valence electron. The valence electron can be in various states with $n = 1, 2, 3, \ldots$; l and $m = 0$. Thus these states have various energies, but no angular momenta, and therefore no magnetic moments. Optical spectra are observed that come from transitions between these "n" states that should not be affected by applied magnetic fields but, in fact, applied fields split the spectral lines into doublets (anomalous Zeeman effect). Also, when beams of these atoms are sent through inhomogeneous magnetic fields they split into two beams (Stern–Gerlach experiment, Born (1989)). In addition to these experiments, according to relativity theory, since the tangential angular velocities of electrons in atoms are very high, the space time coordinate, ct, should be taken into account (c is the velocity of light, t is time). This time coordinate plus x, y, z makes four coordinates. So four quantum numbers might be expected, and in the ideal relativistic four-coordinate theory of P.M. Dirac, the electron itself is found to have quantized angular momentum requiring a fourth quantum number. This fourth number is called the spin quantum number, s. It has only two values, $\pm 1/2$, consistent with the experiments mentioned above and the Dirac theory. So the electron itself has angular momentum $\pm h/4\pi$, and a magnetic moment associated with this, making electrons responsive to applied magnetic fields. This self magnetic moment is in addition to the magnetic moment associated with the orbital angular momentum of the electrons.

The electron's self-moment not only accounts for the anomalous Zeeman effect in atomic spectroscopy, but it has been found to be the magnetic moment that underlies ferromagnetism, paramagnetism, and antiferromagnetism in solids. It is also critical to the theory of superconductivity. Its importance for the theme of this book is through its role in the formation of covalent chemical bonds.

6.6 The Pauli Principle

The one electron in hydrogen seeks the lowest energy state with $n = 1$, if it is not perturbed by an applied field, but what about multi-electron atoms? For example Mg which has an even dozen. Why do not all of the dozen seek to go into the $n = 1$ state? Why, instead, do two of them go into the $n = 1$ state, eight go into the $n = 2$ state, and two go into the $n = 3$ state? Pauli recognized that a simple rule would lead to this structure for Mg, as well as to the observed energy level structures of most of the atoms in the Periodic Table. The rule is simply that *no more than one electron can occupy a single quantum state*. A single quantum state is defined as a state specified by four quantum numbers, n, l, m, and s; and the electronic structure adjusts to minimize the energy of the system. The wave mechanical mathematics of this is summarized in Pauling and Wilson (1983, p. 210).

For Mg, the $n = 1$ state consists of two sub-states with $l = 0, m = 0$, and $s = \pm 1/2$. The $n = 2$ state consists of four sub-states: $n = 2, l = 1, 2,$ or 3 and each having $s = \pm 1/2$. The $n = 3$ state has two sub-states with $l = 0, m = 0, s = \pm 1/2$. The $\pm$ signifies that the electron spins are paired anti-parallel. Pairs of the $++$, or $--$, variety have much higher energy. If only one state of a pair is occupied, an atom has a net magnetic spin moment (magnetic orbital moments can also arise, as well as moments that are vector sums of the orbital and spin moments).

Note that this does not "explain" the electronic structures of atoms. It only *describes* the structure as *deduced* from experimental observations. It is quite remarkable, of course, that such complex structure is contained in the logic of the Schrödinger equation.

6.7 Summing up

The foregoing discussion of electrons in atoms may seem rather detailed in the context of the strength of materials. It has the purpose, however, of showing how a few relatively simple assumptions can lead to complex consequences.

Another purpose is to try to convince the reader that finding short cuts to understanding the behavior of atoms in mechanical systems has not been an easy task. Short cuts have been important, however, since the structures of the solutions of wave equations can be so exceedingly complex.

Also, the reader should be aware that even complex numerical calculations may not give reliable answers to questions that involve the more subtle aspects of the behavior of electrons. Numerical calculations alone yield numbers, but not much insight to the mechanical behavior of materials.

Fortunately, as will be shown, much of the content of quantum mechanics is contained in the Heisenberg Principle (uncertainty).

Finally, this discussion has been a means for introducing nomenclature. This is essential since all of the properties and behaviors of electrons are involved in mechanical behavior at the atomic level. Some play more important roles than others, but none have no role at all.

References

Atkins, P.W. (1983). *Molecular Quantum Mechanics*, 2nd edn. Oxford: Oxford University Press.
Born, M. (1989). *Atomic Physics*, 8th edn., trans. J. Dougall, revised R.J. Blin-Stoyle and J.M. Radcliffe, p. 185. New York: Dover Publications.
de Broglie, M. (1924). *Thesis*, Paris; *Ann. Phys.*, **3** (10) (1925), 22.
Davisson, C.J. and Germer, L.J. (1927). *Nature*, **119**, 558; also, *Phys. Rev.*, **30**, 705.
Pauling, L. and Wilson, E.B. (1983). *Introduction to Quantum Mechanics with Applications to Chemistry*. New York: McGraw-Hill, 1935. Reprinted by Dover Publications, New York.
Vos, M. and McCarthy, I. (1995). Measuring orbitals and bonding in atoms, molecules, and solids, *Am. J. Phys.*, **65** (6), 544.

7

Periodic patterns of electrons

In the last chapter it was indicated that a central Coulomb potential energy field can generate a large number of solutions (probability amplitude functions) of the three-dimensional Schrödinger equation. Quantization of each of the four coordinates, r, θ, ϕ, and t, leads to a large number of possible combinations. However, only relatively few of the possible states are actually occupied by electrons. It is these occupied states that are of interest in interpreting the behavior of various atoms. The occupations of the various states are limited by:

(1) the atomic number which gives the total number of electrons,
(2) Pauli's Principle which limits the number of electrons per state to one,
(3) Hund's Rule which requires the lowest energy states to be occupied first.

The study of optical spectra predated quantum mechanics by many years. In the seventeenth century Isaac Newton discovered that white light passing through a glass prism becomes split into a variety of colors. Early in the nineteenth century, Wollaston observed discrete lines in the light emitted by the sun. This led to observations that individual hot gases emit characteristic series of sharp lines. A dramatic advance occurred in 1885 when Balmer showed that the spectral lines emitted by hydrogen in the visible range are distributed according to a simple formula. The reciprocal wavelength, or wave number, $k = 1/\lambda$, of these lines known as the Balmer series is given by:

$$k = R[(1/4) - (1/n^2)] \qquad \text{where } n = 3, 4, 5, \ldots \qquad (7.1)$$

with R the Rydberg constant, $R = 109\,678 \text{ cm}^{-1}$. Similar distributions were then found for series of lines in the infrared and ultraviolet parts of the electromagnetic spectrum.

The various values of n in this formula correspond to specific (principal) quantum states. The spectral lines with various values of k correspond to transitions between the states with $n = 3, 4, 5, \ldots$ and the state with $n = 2$.

In addition it was observed that the lines of hydrogen had various degrees of sharpness (later, as higher resolution spectrometers became available, it was found that the diffuseness was caused by fine-scale splitting of the lines). Lines of various sharpness were given the following names: s sharp, p principal, d diffuse, and f fundamental. This nomenclature has persisted, and is currently used to describe atomic orbitals. The letters correspond to various values of the angular momentum quantum numbers (orbital shapes):

Symbol	Quantum number l	Shape
s	0	spherical
p	1	dumbbell
d	2	cloverleaf
f	3	cloverleaf (rotated)

Although this double nomenclature does not facilitate learning the subject, it is not likely to change in the near future.

The shapes of orbitals are not limited to those listed above because hybrids can be formed simply by adding orbitals together in particular proportions to form new solutions of the Schrödinger equation; that is, by forming linear combinations of probability amplitude functions in which the coefficients can be determined from the boundary conditions, plus the normalization condition, and the requirement that the functions be orthogonal. The outstanding example is the combination of the s and p functions to form the sp^3 hybrid for carbon atoms where the hybrid has four lobes pointing toward the apices of a tetrahedron. This is the basis of the structure of methane, and of diamond. It is discussed in some detail in Chapter 10.

As has been stated, the principal quantum number, n, determines the total energy of an electron in a quantized orbital (state). However, it also determines the "size" of the orbital. The size is the radius at which the probability density for locating the electron is a maximum (the "mode" of the electron density distribution). For the most simple case, that of hydrogen, this radius is given by:

$$r = (\hbar^2/mq^2)n^2 \tag{7.2}$$

where $\hbar = h/2\pi$, m is the electron mass, and q is the electron charge. In the lowest energy state with $n = 1$, this is equal to 0.529 Å, and is called the Bohr radius, a_0. It is the size of one atomic length unit. For the first excited state with $n = 2$, the modal radius is four times as large.

The energy levels for hydrogen atoms, $E(n)$, are given by the sum of the potential energy, V, and the kinetic energy, T, of the electron in the electric field of the positive nucleus:

$$\begin{aligned} E(n) = V + T &= -q^2/r + mv^2/2 = -q^2(1/r - 1/2r) \\ &= -q^2/2r = -(1/n^2)(q^2/a_0) = -13.6 \text{ eV} \qquad (\text{for } n = 1) \end{aligned} \tag{7.3}$$

This is the energy gained by bringing a widely separated electron and proton together to form a hydrogen atom. Or, it is the ionization energy needed to separate the two particles.

For helium, atomic number 2, the nucleus has a charge of $+2q$, and two orbital electrons. Thus, a first guess might be that the ionization energy ($n = 1$) would be $4 \times 13.6 \, \text{eV} = 54.4 \, \text{eV}$. Wrong! The two electrons interact, mutually screening the nuclear charge from one another, thereby reducing the ionization energy to 24.6 eV. The magnitudes of these electron–electron interactions can be estimated, but not calculated exactly, especially as the atomic number increases beyond two. Thus, although many theoreticians pretend that the subject is deductive, in reality it is highly inductive. There is no set of approximations that can be reliably extrapolated into uncharted territory. Some sets are quite useful for interpolation, however.

With the reader forewarned, we continue to consider atoms with higher atomic numbers. Next in line is lithium with three electrons. To minimize its energy, the tendency would be to put the third electron into the 1s orbital ($n = 1$, $l = 0$) along with the other two. But then, by simple proportionality, since the ionization energies of the first two electrons are 13.6 and 24.6 eV, respectively, this should jump to about 40 eV. Instead, it drops to 5.4 eV. Study of facts like this led Pauli to his daring principle that any particular quantum state cannot be occupied by more than one electron. This seems quite arbitrary, but it has converted so much data from chaos to order that it has come to be accepted as a "law".

There are four quantum numbers: n, l, m, and s. The first three can have large numbers of values (although they usually take on only a few), but s, the self-spin, can only have one of two values, $\pm 1/2$. For two electrons, the spins can either be parallel ($+ +$ or $- -$), or anti-parallel ($+ -$ or $- +$). Thus, according to the Pauli Principle, the 1s orbital of helium can accommodate only two electrons, and the third electron of lithium must go into a higher energy level, namely 2s. Since this is closer to the "vacuum", or zero level, than the 1s levels, the ionization energy is smaller.

It is conventional to list the occupation number as a right-hand superscript of the term symbol, so the complete symbol for lithium is $1\text{s}^2 \, 2\text{s}^1$, and so far we have the following electronic structures:

$$
\begin{array}{llllll}
A = 1 & \text{H} & \rightarrow 1\text{s}^1 & 2\text{s}^0 & \cdots & \cdots \\
2 & \text{He} & \rightarrow 1\text{s}^2 & 2\text{s}^0 & \cdots & \cdots \\
3 & \text{Li} & \rightarrow 1\text{s}^2 & 2\text{s}^1 & 3\text{s}^0 & \cdots \quad \cdots \\
4 & \text{Be} & \rightarrow 1\text{s}^2 & 2\text{s}^2 & 3\text{s}^0 & \cdots \quad \cdots
\end{array}
$$

With $n = 2$, $l = 1$, and $m = -1, 0, +1$ there are three p-states: p_x, p_y, p_z. Therefore, for the next element:

$$
A = 5 \quad \text{B} \rightarrow 1\text{s}^2 \quad 2\text{s}^2 \quad 2\text{p}_x^1 \quad 2\text{p}_y^0 \quad 2\text{p}_z^0 \quad \cdots \quad \cdots
$$

Now Hund's Rule comes into play. The spin of the next electron will interact less with the spin of the one in 2p_x if it uses a new coordinate axis, so it goes into 2p_y (or equivalently, 2p_z).

The pattern for the next few elements becomes:

$$
\begin{array}{llllllll}
A = 6 & \text{C} \rightarrow & 1s^2 & 2s^2 & 2p_x^1 & 2p_y^1 & 2p_z^0 & 3s^0 \ \ldots \ \ldots \\
7 & \text{N} \rightarrow & 1s^2 & 2s^2 & 2p_x^1 & 2p_y^1 & 2p_z^1 & 3s^0 \ \ldots \ \ldots \\
8 & \text{O} \rightarrow & 1s^2 & 2s^2 & 2p_x^2 & 2p_y^1 & 2p_z^1 & 3s^0 \ \ldots \ \ldots \\
9 & \text{F} \rightarrow & 1s^2 & 2s^2 & 2p_x^2 & 2p_y^2 & 2p_z^1 & 3s^0 \ \ldots \ \ldots \\
10 & \text{Ne} \rightarrow & 1s^2 & 2s^2 & 2p_x^2 & 2p_y^2 & 2p_z^2 & 3s^0 \ \ldots \ \ldots
\end{array}
$$

Notice that two "shells" have now been completed. The first, with $n = 1$, is occupied by two electrons and the second, with $n = 2$, is occupied by eight electrons. The next, with $n = 3$, will require $2n^2 = 18$ electrons for its completion. This is known as Stoner's Rule. When the $n = 3$ shell has been completed, we shall begin on the $n = 4$ shell. This will need 32 electrons for its completion.

Perhaps the strongest periodicity in the Table of the Elements occurs each time a shell is completed. Then the next element consists electronically of the completed shell plus one extra electron in the 1s, 2s, 3s, 4s, 5s, ... orbital, respectively. These elements are the alkali metals, H, Li, Na, K, Rb, Cs, ..., and their chemical behaviors are all quite similar. Since they have differing total numbers of electrons, they have different sizes in proportion to the total numbers of electrons. The proportionality is not simply parabolic linear because of the electron–electron interactions mentioned above; the volume per electron (total number) decreases from Li to Cs. However, the atomic volume per valence electron increases on going from Li to Cs. The inverse of this, the valence electron density, plays a key role in determining the elastic stiffnesses and other properties of solids as will be shown in Chapter 12.

When a second electron is added to complete the occupancy allowed in the outermost s-orbital, the chemical behavior is also quite similar with the valence being two instead of one. Thus, the elements Be, Mg, Ca, Sr, Ba, and Ra have similar chemical characteristics.

As more than two electrons are added to the shells, the periodicity becomes less predictable, but does not disappear. Then, as the shells become nearly full, and then completely full, strong periodicity reappears with the halogens, F, Cl, Br, I, and At whose shells have single unoccupied orbitals, and with the noble gases, He, Ne, Ar, Kr, Xe, and Rn whose shells are fully occupied.

It is a triumph of quantum mechanics that the mathematical description of the behavior of probability amplitude waves can account precisely for the behavior of hydrogen atoms, and approximately for the systematic behavior of more than 112 elements.

8

Heisenberg's Principle

Textbooks often emphasize the philosophical aspects of Heisenberg's Principle calling it "The Uncertainty Principle". However, as a physical principle there is nothing uncertain about it. In its "exact" form it is a useful relationship between physical parameters. It is an expression of the fact that particles of matter behave simultaneously in two modes: one that is particle like, and the other that is wave like. These two modes are intimately connected, and it is this connection that is expressed in the Heisenberg Principle, and more generally in quantum mechanics including Schrödinger's equation (Atkins and Friedman, 1997). From this viewpoint, the principle can be used to deduce some of the properties of atoms and molecules in a simple way. This method of analysis is quantitative, but it is limited to cases that involve only a single coordinate that is given in advance (Born, 1989).

The method has been extended somewhat by Simons and Bloch (1973) to cover some cases involving the angular momentum quantum number. It is a rudimentary form of density functional theory. In a somewhat different form than the one used here, it has been applied to molecules by Parr and his collaborators (Borkman and Parr, 1968).

Physical properties can never be measured exactly. Errors, however small, are always present in a measurement procedure. Therefore, measured values of variables are average values derived from a distribution of individual measurements. The spread of the distribution (assuming it is symmetric) is conveniently described in terms of its second moment, that is, in terms of the standard deviation of the distribution from the average value. For some measurements, the individual values cluster close to the average value, so the standard deviation is small compared with the average. For other measurements, the values are widely spread, and their standard deviation is large.

Let the quantity being measured be the position x_i (one dimension) of an electron. Suppose the N measurements of the position are made. Then the average value, $\langle x \rangle$, of the position is:

$$\langle x \rangle = \frac{1}{N} \sum_i^N x_i \tag{8.1}$$

The deviation of a particular position x_j from the average is $(x_j - \langle x \rangle)$, so the average of

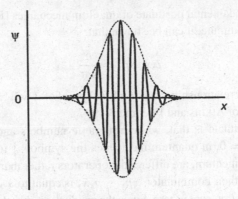

Figure 8.1 Pulse formed by modulating a carrier wave. The carrier (solid line) has a wavelength λ, and the modulation envelope (dotted line) is Gaussian.

the squares of the deviation is:

$$\left(\frac{1}{n}\right) \sum_{i=1}^{n} (x_i - \langle x \rangle)^2 = (\Delta x)^2 \tag{8.2}$$

where Δx is the standard deviation from the average. Note that x_i can be either plus or minus.

Now consider an electron (or other particle) to be moving along the x-axis (Figure 8.1). At an instant of time, it has a pulse of probability amplitude (wave function) associated with it that has a wavelength, λ, and a modulation envelope (the dotted line in the figure). It also has a rest mass m. Then, according to the de Broglie hypothesis, as verified by experiments (electron diffraction, etc.), its momentum is $p_x = h/2\pi\lambda$. The shorter is its wavelength, the higher is its frequency; and therefore the higher is its energy according to Planck's hypothesis, $E = h\nu$.

To localize the particle, the carrier wave (probability amplitude) is multiplied by a modulation pulse (of unit amplitude). This is the dotted line in Figure 8.1. Since the description is probabilistic, the shape of the modulation pulse is taken to be the Gauss error function $\exp(-x^2/2)$. To simplify the algebra, a "snapshot" is taken at time $t = 0$, with the pulse center at $x = 0$. The standard deviation of the position, Δx, is defined such that it is the half-width of the amplitude modulation pulse, equal to $1/\sqrt{2}$.

The pulse modulation envelope affects the local wavelength of the carrier wave (that is, it affects the momentum, or kinetic energy, of the particle). It does this by shifting the peaks, and therefore the valleys, of the carrier wave inside the modulation envelope. If Δx becomes smaller, the distances between the peaks become smaller, so the wavelength decreases. The extent of the variation of the wavelength, λ, is its standard deviation, $\Delta\lambda$. And, since $p_x = h/2\pi\lambda$, the variation of λ causes a variation of p_x, the standard deviation being Δp_x.

As a result of the fundamental postulate of quantum mechanics (Heisenberg's postulate) that x and p_x do not commute, it can be shown that:

$$\Delta x \ \Delta p_x = \frac{h}{4\pi} \tag{8.3}$$

This is a very important relationship, so a simplified proof of it will be presented. This proof follows the discussion of Atkins and Friedman (1997).

The Heisenberg postulate is that, while arithmetic numbers and their representations commute, so $ab - ba = 0$, in quantum mechanics the symbols x for position, and p_x for the x-component of momentum, are differential operators rather than numbers, and they do not commute. Instead, their commutator, $xp_x - p_xx$, is equal to a very small constant. In classical mechanics this constant is zero. Since the product xp_x has the dimensions of action (energy times time) the small constant was chosen to be close to Planck's action constant, h, which is small but not zero. This choice was then shown to be in agreement with a large number of experiments. The non-commutation rule became:

$$xp_x - p_xx = ih/2\pi$$

where $i^2 = -1$.

Some operators do commute. For example, if x and y are simply algebraic variables, then $xy - yx = 0$. However, if x and y are operators with x meaning "multiply by x", and y meaning "differentiate with respect to x" (that is, $y = d/dx$), then, for a function such as $z = f(x) = x^3$, when the operations are applied in turn from right to left, $xy[z] = 3x^3$ but, if they are applied from left to right, $yx[z] = 4x^3$; so the operators do not commute, and $xy - yx = -x^3$.

To connect this with physical effects an example can be taken from vector algebra. Vectors are operators, and do not commute. Consider a rabbit whose position and posture are determined by three orthogonal vectors, **a**, **b**, **c**. Let **a** point forward, **b** point to the right, and **c** point upwards. The rabbit starts out looking forward along **a** with ears parallel to **c**. Define a vector operation **x** which means "turn 90° clockwise (looking toward the origin) around the vector, **x′**." Then the sequence **x**, **x′** = **b**, **a** means "turn the rabbit on its back, looking upwards; then turn the rabbit on its back so the ears point to the right." However, the sequence **a**, **b** puts the rabbit on its side with ears pointing to the right, and then rotates the rabbit on its side until it looks to the right with its ears pointing backward. Clearly **b**, **a** is not equal to **a**, **b**. Thus vector operations (three-component matrices) do not commute in general, nor do larger matrices. This is why Heisenberg's version of quantum mechanics is called matrix mechanics.

Equation (8.3) is very useful because it contains some of the essential features of quantum mechanics while finessing the complexities. It states that the standard deviation of the position times the standard deviation of the momentum (kinetic energy) equals a known constant. Therefore, if common sense yields a reasonable distribution for the position of a particle, and the standard deviation of this distribution can be found, then the standard deviation of the kinetic energy is given without a need for solving a difficult differential equation. This is not as satisfying as finding the exact solution for a problem, but it is often as good an approximation to the solution as any other, and it is almost always a good method for estimating solutions. It will be used later in this book in connection with elastic stiffness.

This approach is approximately equivalent to density functional theory. It yields energy values for the lowest energy states, particularly if the size of the system is assumed to be known from experimental data. Since the sizes of most systems of interest to the science of materials are in fact known from diffraction experiments, this considerably simplifies calculations, and yields rather good results. Also, it indicates why many of the mechanical properties of interest depend primarily on electron densities, that is, the number of electrons in a local unit volume. Furthermore, it is very similar to the "bond charge model" of Borkman and Parr (1968) which has been successfully applied to a variety of molecules.

The Heisenberg Principle is entirely general. That is, it holds for all physical situations. However, since h is very small, the constraint imposed by Equation (8.3) on macroscopic situations is often negligible. Consider an automobile that weighs 1000 kg and is moving at 27 m/s (approximately 60 mph). According to Equation (8.3) the limit with which its position can be measured (with $h = 6.63 \times 10^{-34}$ J s) is about 3×10^{-40} cm, which is negligible indeed! On the other hand suppose an electron has a kinetic energy of 10 eV = 16×10^{-12} erg. Then the minimum standard deviation of its position is 0.31 Å, or about one-tenth the size of an atom.

To prove Equation (8.3), the Gaussian form for the pulse envelope will be used; $\exp(-x^2/2)$. This modulates a sine wave, $y = \cos(2\pi x/\lambda) = \cos k_x x$, where λ is the wavelength. The wave number, $k_x = 2\pi/\lambda$, determines the momentum, $p_x = hk_x/2\pi$, of the packet. Figure 8.1 illustrates a packet with average carrier wavelength of $\sqrt{2}$ atomic units, equal to 0.75 Å (one atomic unit is equal to the Bohr radius, 0.53 Å). The Gaussian envelope of the packet is centered on $x = 0$, so its average, $\langle x \rangle = 0$, but its standard deviation, $\Delta x = \langle x^2 \rangle \neq 0$. Similarly, $p_x = -p_x$, so $\langle p_x \rangle = 0$, but the standard deviation of the momentum, $\Delta p_x = \langle p_x^2 \rangle \neq 0$. The square of the standard deviation of the momentum is proportional to the average value of the kinetic energy, T, since:

$$T = p_x^2/2m = h\omega/2\pi$$

Figure 8.1 indicates that as the width of the envelope of the wave packet increases (i.e., Δx increases), the carrier wavelength that it can accommodate increases. Conversely, if Δx decreases, the wavelength decreases. In classical mechanics the wavelength spread is independent of the position spread, but this is not the case for electrons which have the dual characteristics of particles and waves. For an electron, its position is linked to its momentum (kinetic energy) because the product $\Delta x \Delta p_x$ is equal to a constant. Since standard deviations are commonly associated with Gaussian distributions, it is convenient to use the Gauss error-function for the probability amplitude, ψ, of the electron (it must be normalized so the total probability for the electron to be somewhere equals unity).

$$\psi = \left(\frac{1}{\pi}\right)^{1/4} e^{-x^2/2} \tag{8.4}$$

Then, the mean square value of x is:

$$\langle x^2 \rangle = \int_{-\infty}^{\infty} \psi x^2 \psi^* \, dx = (1/\pi)^{1/2} \int_{-\infty}^{\infty} x^2 e^{-x^2} \, dx = 1/2 \tag{8.5}$$

while the mean square momentum is:

$$\langle p_x^2 \rangle = (1/\pi)^{1/2} \int_{-\infty}^{\infty} \psi \left[-(h/2\pi)^2 \frac{d^2}{dx^2} \right] \psi^* \, dx = 1/2(h/2\pi)^2 \tag{8.6}$$

with the term in square brackets being the momentum operator. Thus the product of the standard deviations is:

$$\Delta x \, \Delta p_x = \left[\langle x^2 \rangle \langle p_x^2 \rangle \right]^{1/2} = \frac{h}{4\pi} \tag{8.7}$$

and similar expressions hold for the y and z coordinates. Accordingly, if an electron is confined by a potential well to a very small range of x, its momentum (kinetic energy) must increase to satisfy Equation (8.7). This has been confirmed in a variety of micro-electronic devices.

Conversely, if an electron can spread out in space, it can reduce its momentum. This is the principle underlying the formation of chemical bonds. When two, or more, atoms come into contact and their energy levels match, there is a possibility for their valence electrons to spread over the total volume, thereby decreasing their energies and bonding the aggregate together.

This principle also underlies elastic stiffness. If an atom is squeezed, the confinement of its electrons is increased so their kinetic energy increases causing resistance to the squeezing.

8.1 Heisenberg hydrogen atom

A picture of the hydrogen atom in terms of an exact solution of the Schrödinger equation has already been described. A simplified approach uses the Heisenberg Principle. For a one-electron hydrogen-like atom it contains the same physical ideas as the Schrödinger approach, but is limited to the lowest energy state. Its simplicity recommends it.

Consider a positive proton immersed in a charge cloud formed by one valence electron. The cloud is of uniform density, and is spherical with a radius R. This radius is determined by a balance between: (1) the electrostatic attraction between the positive proton and the negative electronic charge; and (2) the effect that confining the electronic charge to a region of radius R has on its kinetic energy. The former effect tends to contract the atom, while the latter tends to expand it. The former is given by standard electrostatic theory, and the latter by the Heisenberg Principle.

To calculate the kinetic energy of the electron cloud, since momentum is a linear vector, it is necessary to begin by expressing the mean square radius $\langle r^2 \rangle$ in terms of its x, y, and z components:

$$\langle r^2 \rangle = \langle x^2 \rangle + \langle y^2 \rangle + \langle z^2 \rangle = 3\langle x^2 \rangle$$

and, since $\langle x \rangle^2 = 0$ for a sphere, the standard deviation of x is given by:

$$(\Delta x)^2 = \langle x^2 \rangle - \langle x \rangle^2 = \langle x^2 \rangle$$

with similar expressions for the y and z coordinates. Applying the Heisenberg Principle:

$$\langle r^2 \rangle = 3(\Delta x)^2 = 3(h/4\pi \Delta p_x)^2$$

For a sphere, the averages of the linear momenta are zero. That is:

$$\langle p_x \rangle = \langle p_y \rangle = \langle p_z \rangle = 0$$

Therefore:

$$(\Delta p_x)^2 = \langle p_x^2 \rangle = (3/16)(h/\pi r)^2$$

with similar expressions for the y and z components; and the expression for the kinetic energy becomes:

$$T = (1/2m)\left[3\langle p_x^2 \rangle\right] = 9h^2/32\pi^2 m r^2 \tag{8.8}$$

where m is the electron mass. This shows that T is inversely proportional to r^2, so the more the electron is confined the greater is its kinetic energy. Inserting the constants, h and m, yields $T_H = 2.25/r^2$ which compares well with the value given by the free-electron theory using wave functions; namely, $T_E = 2.21/r^2$ (the difference is less than 2%). It should be kept in mind that since the Heisenberg Principle relates minimum standard deviations, the expression for T gives the minimum energy for the given configuration.

The electrostatic potential energy, under the approximation that the charge cloud is of uniform density from the origin to the radius, R, can be found by calculating the electrostatic potential from $r = 0$ to $r = \infty$, and then calculating the work done in bringing an electron from infinity to any position, say $r = R$. For any radius r the electric field seen by a test charge placed at r is zero for $r = 0$, increases linearly with r until $r = R$, and then decreases inversely with increasing r. This results from the Gauss theorem which states that the field seen by a test charge placed outside of a region containing distributed charge is the same as if the distributed charge were located entirely at $r = 0$.

Thus, outside of the spherical charge cloud, the electric field $E = q/r^2$, where $r > R$ and q is the charge of the electron. For points inside the cloud ($0 < r < R$), the field results from the charge that lies inside a sphere of radius r as if all the charge lying at a greater distance were removed. In this region, $E(r) = (q/r^2)(r/R)^3 = qr/R^3$, where $r < R$. Knowing the field, the potential energy, P, can be obtained by integration. There are two cases of interest: P_{pp} the point–point, and P_{pc} the point–cloud energies

$$P_{pp} = P_{pc} = \int_R^\infty E \, dr = q/r \qquad\qquad (r \geq R)$$

$$P_{pc} = \int_r^R E \, dr = (q/R)\left(\frac{3R^2 - r^2}{2R^2}\right) \qquad (r \leq R)$$

Since the proton is at $r = 0$, for the hydrogen atom $P_{pc} = 3q/2r$. The total energy is the sum of the kinetic and potential energies:

$$E(r) = T + V_{pc} = [(9h^2)/(32\pi^2 m r^2)] - [(3q^2)/(2r)]$$

This can be minimized by setting the first derivative equal to zero, $dE/dr = 0$, and solving for the equilibrium value, $r = r_0$. This gives: $r_0' = 3h^2/8\pi^2 m q^2$ (cgs units). If this is substituted back into the expression for $E(r)$, it yields $E(r_0) = -2\pi^2 m q^4/h^2 = -13.59$ eV, which is exactly the experimental value. This precision is coincidental, however, since the value for r_0 that the model gives is 50% larger than the correct Bohr radius, $a_0 = h^2/4\pi^2 m q^2$, but it shows that we can begin to understand the structure of atoms through the use of just two laws: one from de Broglie and Heisenberg, and one from Coulomb.

As mentioned earlier, the Heisenberg model only deals with the lowest energy state of the hydrogen atom. To discuss the higher energy states, the Schrödinger equation must be used.

References

Atkins, P.W. and Friedman, R.S. (1997). *Molecular Quantum Mechanics*, 3rd edn. Oxford: Oxford University Press.

Borkman, R.F. and Parr, R.G. (1968). Toward an understanding of potential-energy curves for diatomic molecules, *J. Chem. Phys.*, **48**, 1116.

Born, M. (1989). *Atomic Physics*, 8th edn., trans. J. Dougall, revised R.J. Blin-Stoyle and J.M. Radcliffe. New York: Dover Publications.

Simons, G. and Bloch, A.N. (1973). *Phys. Rev. B*, **7**, 2754.

Section IV
Elastic stiffness

9
Cohesion of atoms

The ideas involved in the theory of cohesion which underlies the theory of strength are relatively simple, although the details may be intricate. The most important forces that give high cohesion, and therefore high strength, are electrostatic so the theory might be said to have begun with Benjamin Franklin (1752). He speculated, based on his studies of electrostatics, that charges were somehow involved in cohesion. Later, Berzelius (1819) proposed that atomic particles with net positive and negative charges attract one another thereby forming chemical bonds. However, this did not account for the binding between atomic particles of the same electrostatic sign.

Verification of these speculations came only after a century of advances in technology. Methods for conveniently generating, storing, and handling electricity had to be developed as well as improved methods for evacuating containers. During his search for effective incandescent filaments, the improved vacua that he developed led Thomas Edison (1884) to the discovery that hot conductive filaments emit electricity that can flow to positive collectors through vacua. One of the consequences of this discovery, occurring about ten years later, was that Wilhelm Roentgen (1895) discovered X-rays.

Although atomic theories of matter had been postulated for millennia, and estimates of atomic spacings and structures in crystals could be made on the basis of the work of Boyle, Dalton, and Faraday, definitive measurements did not exist. Then, 17 years after Roentgen's discovery, Max von Laue (1912) showed that X-rays can be diffracted by crystals. This allowed the arrangements and spacings of the atoms in crystals to be determined very accurately. Typical spacings are about 0.2 nm (2 Å).

By this time J.J. Thomson had shown that Edison's cathode rays consisted of discrete particles (electrons), and had determined the masses and approximate charges of the electrons.

Now a substantial theory of cohesion began to emerge. The force between a positive and negative particle spaced 0.2 nm apart, and each having one electronic charge, would be (q^2/r^2) according to Coulomb's Law, where q is the electronic charge, and r is the spacing.

63

Dividing this by an atomic area, r^2, makes it a stress (energy per unit volume) equal to 150 GPa. This equals a typically measured elastic modulus, so the idea that electrostatic forces can account for cohesive forces became quantitatively justified. By adding up the attractive and repulsive electrostatic interactions between the ions in salt crystals a good account of the cohesive energies could be made, and it was found that they compare quite well with values determined by thermochemistry. However, the theory of ionic crystals did not account for covalent or metallic bonding.

The great idea that pushed the theory of atoms from the realm of physics into that of chemistry was that of de Broglie (1924). His postulate of the wave-like character of electrons provided a means for rationalizing various patterns of electron density in molecules and crystals, including calculations of the strengths of the electron-pair bonds between neutral atoms. G.N. Lewis had qualitatively proposed these pair bonds in about 1916. Particles with mass, m, such as electrons, moving at a velocity, v (in a space free of electric and magnetic fields), have momentum $p = mv$, and kinetic energy $T = mv^2/2 = p^2/2m$. Then, from Planck's theory, $T = h\nu$, where h is Planck's constant, ν is frequency, and if the electron is allowed to be relativistic, $v \to c$, and $T \to mc^2$, Thus, since $c = \lambda\nu$:

$$p = h/\lambda \qquad \text{de Broglie's postulate}$$

As discussed in Chapter 5, this immediately rationalized the nagging question of why the momenta of orbiting electrons (in the Bohr theory of the atom) are quantized.

The bonding that occurs between atoms, within molecules, and between molecules, results from the interplay between the charged particle aspect of electrons, the wave-like aspect, and the constraints placed on the distribution of charge by the Pauli Principle. After the distribution of charge within and between the atoms of a material has been determined by these factors, the energy and forces can be determined by means of standard electrostatic theory according to a theorem proved by Hellman and Feynman (see Weiner, 1983). Thus, after the charge distributions have been modulated by the rules of quantum mechanics, the fundamental forces are electrostatic. These forces, when they are distributed over surfaces and volumes, become expressed as elastic stiffnesses.

Since electrostatic forces vary inversely with the square of the distance between charged elements, a key parameter in the theory of bonding is the sizes of atoms (i.e., bond lengths). These can be found by means of X-ray, electron, and neutron diffraction techniques. The spacings are approximately 0.2 nm; the exact values depend on the particular atoms. The distributions of electronic charge can be verified experimentally from X-ray scattering (Coppens, 1997). At one time X-ray scattering gave good values for the spacings of atoms, but little information about the distribution patterns of the valence electrons. However, this is no longer the case. In addition, from particle scattering, the radii of atomic nuclei have been determined to be 1.5–30 fermis (1 fm $= 10^{-13}$ cm), that is, about 10^5 times smaller than whole atoms. Thus, the positive nuclear charge can be taken to be concentrated at the very centers of atoms, to a very good approximation. Furthermore, since the positive nuclei are 1800, or more, times heavier than the electrons that swarm about them, they can almost always be taken to be stationary relative to the electronic motions. A corollary of this is that during a structural change, the electronic structure is always approximately in equilibrium with the instantaneous positions of the nuclei. Therefore, the electronic structure leads while the nuclear arrangement follows during a structural change.

In recent years even the energy (momentum) distributions of electrons in atoms, molecules, and crystals can be determined experimentally. This is done by means of co-incidence detection of incident and scattered electrons, which allows the momenta of the target electrons to be determined (see Figure 6.6).

By taking bond lengths and crystal structures as given by experimental crystallography, much less computation is needed to identify the factors that determine cohesion. This can then be applied to gain improved understanding of the mechanical properties and processes. As will be shown, the key parameter for determining elastic stiffnesses and surface energies is valence electron density. Therefore, if the atomic volume is taken from crystallography, theory only needs to supply the number of valence electrons. Since the principal interest in this book is not the theory of electrons, but the theory of mechanical behavior, using atomic sizes as inputs is quite justified.

Although bonding is primarily a result of electro*static* forces, there is an important component that comes from electro*dynamic* fluctuations (hopping of small particles between heavier ones). For electrons, this was originally suggested to be important by Heisenberg (1930). In materials there are four principal cases (for nuclei there are additional cases):

(1) electrons hopping between ions (covalent bonds),
(2) photons hopping between molecules, or atoms (London and Casimir bonds),
(3) protons hopping between atoms (hydrogen bonds),
(4) phonons hopping between conduction electrons (superconductivity).

The hopping effect is analogous with the behavior of coupled pendulums (or other oscillators). For a system of two pendulums, the coupling produces two limiting states: one in which the pendulums move in synchronism, and the other in which they move anti-synchronously. The former has the lowest energy for the system so it corresponds to bonding, while the latter higher energy state corresponds to anti-bonding.

By forming time averages, these dynamical models can be converted into charge distributions and treated by the methods of electrostatics.

Dynamics is critical to understanding the cohesion of neutral atoms (rare gases), and neutral molecules (polymer molecules), as well as the Casimir–London forces between macroscopic pieces of material, and the van der Waals forces in non-ideal gases. According to quantum electrodynamics, the charges within neutral entities are always in states of oscillation (Craig and Thirunamachandran, 1998). This is a consequence of the wave mechanics of harmonic oscillators. If their oscillation frequency is ν, their lowest energy state is $h\nu/2$ (the zero-point energy). This is consistent with Heisenberg's Principle which states that a particle localized to an oscillator of size Δx has kinetic energy not less than $(1/2m)(h/4\pi\Delta x)^2$, where m is mass, and h is Planck's constant. Thus, even at $T = 0$ K, charge oscillators (dipoles) are active, and attractive to one another.

A complication arises for atoms with large atomic numbers (large nuclear charges). In this case, the effective velocities of the electrons that are pulled close to the nuclei become very high. Therefore, their energies need to be corrected by the requirements of the special theory of relativity. The corrections are usually small so they can be neglected, but they do affect chemical reactions of the platinum group metals (Bond, 2000).

Detailed electronic charge distributions are not easily determined from quantum mechanics. Too many variables affect them. Numerical methods using the fastest of computers give only good, but not entirely reliable, results. Nevertheless, it is remarkable how much can be accounted for by modern theory, especially in terms of relatively simple models that have been developed over many years, some by chemists and some by physicists. Simplified models became possible as the structure of wave mechanics became clarified, so that simple but effective approximations could be selected.

9.1 Limiting bond types

There are three limiting types of strong chemical bonds (Figure 9.1): metallic, ionic, and covalent. They correspond to the three possible distribution patterns for charge in the vicinity of two, or more, atomic ions. The figure places these three pattern types at the vertices of a triangle, following Van Arkel and Ketelaar (1956). Some intermediate types lie along the edges.

The limiting types of strong bonds have the charge distribution patterns shown schematically in Figure 9.2. In the covalent case, the charge associated with electron pairs becomes localized between the ions. In the ionic case, one or more electrons can lower the energy of a system by transferring from one ion to the other of a pair where it is localized. It then attracts the positive ion left behind. Finally, in metals the bonding electrons are delocalized, so they are distributed uniformly with the positive ions immersed in the liquid-like distribution of electrons.

Filamentary or layered structures involve a fourth type of bonding that is weak compared with the three limiting types. It is associated with the interactions of charge dipoles, instead of monopoles. It acts, for example, in polymers such as polyethylene. The polymer chains are bonded internally by bonds of the covalent type between carbon (or other) atoms, but the chains are held together by much weaker dipole–dipole bonds. A result is that a random array of the polymeric molecules is soft as in polyethylene bottles. On the other hand, fibers in which the molecules are all aligned in one direction are very stiff as in Spectra fibers (highly oriented polyethylene).

Hydrogen bonding is a fifth type that is sometimes important when water with its two hydrogen atoms is present, particularly in biological systems.

In Figure 9.1 most substances lie within the triangle which implies that most bonding consists of mixtures of the ideal types. That is, they are hybrids. Some of these hybrids lie close to the edges of the figure's triangle, but most of them lie somewhere in the interior.

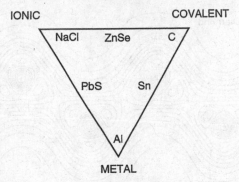

Figure 9.1 Diagram relating solids with the three principal types of chemical bonding. Examples of the pure types are given at the apices of the triangle; a few intermediate types are shown along the edges. Note that elements lie along the monoatomic metal–covalent line, whereas diatomic compounds lie along the other axes. Inside the triangle are mixed types.

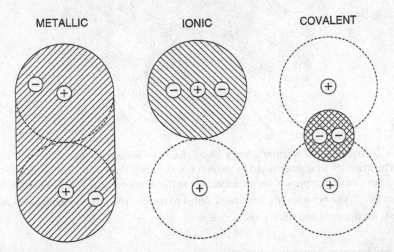

Figure 9.2 Schematic sketches of the electron distributions for the three principal bonding types. Left, metallic with positive nuclei immersed in valence electron "liquid"; middle, ionic with valence electron transferred from one atom of a pair to the other; right, covalent with localized electron pair between two positive ions.

9.2 Covalent bonds

In some covalently bound solids, such as Ge, Si, and C crystals, the pattern of valence charge has been determined from X-ray scattering measurements (carbon, Spackman, 1991; silicon, Deutsch, 1992; germanium, Brown and Spackman, 1990). Figure 9.3 shows the measured electron distribution for C–C bonds in diamond, and Si–Si bonds in silicon. This distribution can also be calculated numerically from approximate theory (Lu and Zunger, 1993). It is of special interest here because it does not conform to the elementary theory of molecular

Figure 9.3 Distributions of electron density along chains of bonded atoms in diamond and silicon crystals. The densities were determined by means of X-ray scattering. Note the bimodal distributions along the lines connecting pairs of atoms, especially in the diamond case. The positions of the atoms are marked by +. The bond lengths have been scaled to match. The actual bond lengths are 1.54 Å and 2.35 Å for diamond and silicon, respectively.

orbitals; and because it does account for the exceptionally large shear stiffness of diamond. It may be seen that the charge distribution is bimodal, instead of being unimodal as elementary theory suggests (Gray, 1973). That is, along the lines that connect pairs of atoms, there are two peaks of charge lying at the 1/4 and 3/4 positions along the bond.

Figure 9.4 shows a charge (electron) density map for a layer of the graphite structure (Stewart and Spackman, 1981). The hexagonal pattern of atoms is apparent, and again the distribution of electrons along the bonds is bimodal with peaks at the 1/4 and 3/4 positions. In this case the bond length is 1.42 Å, so the bond charge density is roughly 8% greater than it is in diamond. This is consistent with the very high elastic stiffness of graphite parallel to its layers.

Knowing the distance, d, between two atoms, and that the valence charge is concentrated between them, it is then a simple matter to estimate the cohesive energy per bond from

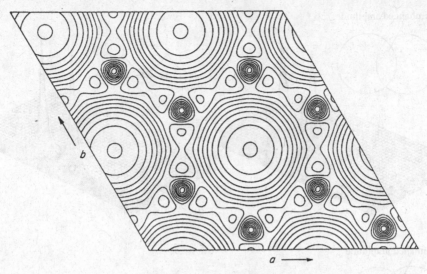

Figure 9.4 Electron density map of a layer of graphite.

Coulomb's Law. If q is the electron's charge, the cohesive energy is about $(3/8)(q^2/d) = 3.5\,\mathrm{eV}$ for carbon. This may be compared with the measured value of $3.6\,\mathrm{eV}$ (the $3/8$ factor is an estimate of the effect of the screening by the core electrons on the valence electron energy). This is known as the *bond charge model* (Parr and Yang, 1989).

In covalent crystals, spin-pairing just saturates the bonding molecular levels. If there are additional electrons in a particular system they must occupy higher energy non-bonding, or anti-bonding, molecular levels. Unlike the localization of the bonding levels between the atoms, the anti-bonding levels allow the electrons to become delocalized, so they spread out, and are able to conduct electricity. The result is a semiconductor, or a metal if the number of excess electrons is large.

9.3 The importance of symmetry factors

For simple metallic and ionic crystals the charge distributions on the atoms are approximately spherical, but this is not the case for other metals, nor for covalent crystals. Also, arrays of atoms may, or may not, have spherical symmetry. Perhaps the most important cases for mechanical behavior are the arrays of atoms at the cores of dislocations. In this case the symmetry changes as material passes into, and then out of, the core of a moving dislocation. This will be discussed in some detail later (Chapter 18). Here, the effect of symmetry on bonding is very important, in particular the effect of symmetry on the formation of bonding molecular orbitals.

For an LCAO (linear combination of atomic orbitals) to be effective in forming a bond, the two atomic amplitude functions, ψ_A and ψ_B, need to have particular characteristics (Coulson, 1952). Thus, not all chemical elements combine to form compounds. The first

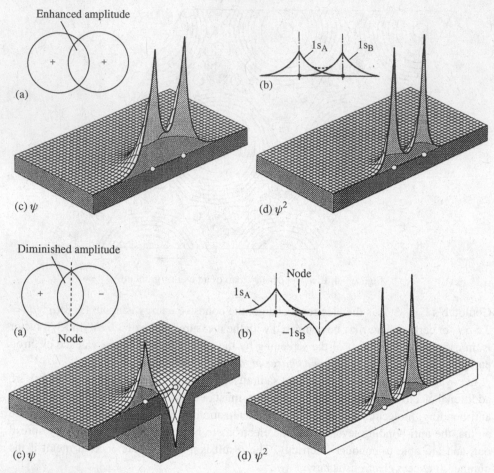

Figure 9.5 Effects of overlapping probability amplitude functions. Top, in-phase bonding overlap; bottom, out-of-phase anti-bonding overlap.

characteristic is that they must have similar energies, that is, similar wavelengths, according to de Broglie's postulate. Otherwise, when the probability amplitude of the molecular orbital is squared and integrated over all space, the result will be zero, indicating that there are no valence electrons in the region, and no bonding. This can be seen quite simply for periodic functions. If one of them has a long wavelength, and the other a short wavelength, the interaction term, $\psi_A\psi_B$, has just as many positive as negative regions so its integral vanishes, and only the atomic terms, $|\psi_A|^2$ and $|\psi_B|^2$, are left, so there is no bonding.

The second characteristic is that the atomic amplitudes, ψ_A and ψ_B, must overlap as much as possible. It is self evident that there must be overlap; otherwise there would be no interaction. That the effect is monotonic can be seen as follows for the most simple case, that of diatomic molecules. This illustrates the principles (see Figure 9.5 (top)).

Take two atoms, A and B (Figure 9.5(top a)). Let the probability amplitudes of their valence electrons, when they are well separated, be ψ_A and ψ_B, respectively. Assume they are each normalized, so: $\int |\psi_A|^2 \, dv = \int |\psi_B|^2 \, dv = 1$. Form a molecular orbital ϕ_{AB} from the atomic probability amplitudes by adding them together linearly (i.e., make an LCAO, Figure 9.5(top b)):

$$\phi_{AB} = a\psi_A + b\psi_B \qquad (9.1)$$

where the coefficients a and b give the fractional portions of ψ_A and ψ_B. If $A = B$, $a = b$. If $A \neq B$, a and b are determined by minimizing the energy, U, of the system, that is, by minimizing:

$$U = \int \phi \mathbf{H} \phi \, dv \bigg/ \int \phi^2 \, dv \qquad (9.2)$$

where $\mathbf{H}$ is the energy operator, and $\int dv$ means integration over all space (i.e., over all of the coordinates of the system). Equation (9.2) is obtained from the Schrödinger equation:

$$\mathbf{H}\phi = U\phi \qquad (9.3)$$

Multiplying both sides of Equation (9.3) by ϕ (remembering that $\mathbf{H}$ is an operator, while U is a number) yields:

$$\phi \mathbf{H} \phi = U \phi^2 \qquad (9.4)$$

Integrating both sides of this and solving for U gives Equation (9.2). This is plotted schematically in Figure 9.5(top c). The probability amplitude of the molecular orbital of Figure 9.5(top c) is squared to obtain the probability density of Figure 9.5(top d).

A powerful principle is the Variation Principle (Rayleigh–Ritz Principle) as it applies to Equation (9.2). According to this principle, the solution of Equation (9.3) when substituted into (9.2) produces the minimum value of U (Atkins, 1991). Therefore, if any reasonable guess for the function, ϕ, is put into Equation (9.2), the resulting U will be greater than, or equal to, the correct energy. Then, if U is known from experiment, and the calculated U is equal to the experimental value, the ϕ function that has been guessed is the correct probability amplitude. Or, if the guessed function has a parameter in it that can be varied, U can be determined for various values of the parameter until U_{min} is found. The corresponding ϕ is then the correct one. If the guessed function has more than one parameter then U must be minimized with respect to all of the parameters simultaneously.

The effect of overlap may be seen by substituting the molecular orbital into Equation (9.2):

$$U = \int \phi_{AB} \mathbf{H} \phi_{AB} \, dv \bigg/ \int \phi_{AB}^2 \, dv \qquad (9.5)$$

where $\phi_{AB}^2 = a^2 \psi_A^2 + b^2 \psi_B^2 + 2ab\psi_A \psi_B$, and the third term in this expression is the interaction term that describes the distribution of charge between the two atoms. The algebra can be considerably simplified by switching to a homonuclear molecule so that $A = B$, $\psi_A = \psi_B$, and $a^2 = b^2$. Thus $a = \pm b$, and there are two possible molecular orbitals (just equal to the number of atomic orbitals in the system).

The atomic orbitals are normalized, so $\int \psi_A^2 \, dv = \int \psi_B^2 \, dv = 1$, but the molecular orbitals, $\phi_{AB} = N_+(\psi_A + \psi_B)$ and $\phi_{BA} = N_-(\psi_A - \psi_B)$, have not yet been normalized. N_+ and N_- are the normalization factors. The normalization condition for the orbital ϕ_{AB} is $\int \phi_{AB}^2 \, dv = 1$, and since the atomic orbitals have been assumed to be individually normalized, for this molecular orbital:

$$N_+^2[2(1 + S)] = 1$$

where S is the overlap integral, $S = \int \psi_A \psi_B \, dv$; similarly

$$N_-^2[2(1 - S)] = 1$$

For large separations of the atoms, S approaches zero, and it approaches unity for complete overlap of the ψ.

With its normalization factor, the amplitude for the bonding molecular orbital can be written:

$$\phi_+ = (\psi_A + \psi_B)\{1/[2(1 + S)]\}^{1/2} \tag{9.6}$$

and it may be seen that the larger the overlap, the smaller the energy, and the stronger the bonding. In minimizing the total energy, the repulsion of the positive nuclei must be considered. This limits the amount of overlap.

The case of anti-bonding is illustrated in Figure 9.5 (bottom). Here the amplitude for the anti-bonding orbital is:

$$\phi_- = (\psi_A - \psi_B)\{1/[2(1 - S)]\}^{1/2} \tag{9.7}$$

so increasing the overlap, S, increases the energy in this case. Therefore, in addition to the repulsion of the two positive nuclei, electronic repulsion also occurs. This is sometimes called Pauli repulsion.

There is another condition that supercedes overlap. This is the third characteristic that determines whether strong bonding occurs. It is a symmetry condition that requires the overlapping amplitude functions to have the same symmetries, otherwise bonding does not occur. A simple example is the case of an s-function on one atom, and a p-function on the other. Let the potential bonding coordinate be the x-axis, and the atom with the spherical s-function be located at the center of the xyz coordinate system. Then there are three possibilities for the orientation of the approaching dumbbell shaped p-type atom. They are p_{xx}, p_{xy}, and p_{xz}. In all three cases, the center of the dumbbell lies on the x-axis, while the axis of the dumbbell lies along the x, y, or z axis respectively. In the xx case, the p-function presents cylindrical symmetry to the cylindrical symmetry of the s-function (with respect to the x-axis) and bonding can occur. However, in the xy and xz cases, the dumbbell presents anti-mirror symmetry to the mirror symmetry of the s-function, and bonding cannot occur. In these latter cases, if the phase of one of the lobes of the p-function matches the phase of the approaching s-function, the other lobe is necessarily out of phase, so cancellation occurs and there can be no bonding.

In summary, the conditions on the participating quantum states for significant bonding to occur are:

(1) approximately equal energies,
(2) appreciable overlap of the amplitudes of the two states,
(3) compatible symmetries of the amplitudes of the two states.

These conditions have been described in some detail here because they are critically important to an understanding of the mobility of dislocations, and therefore to an understanding of plastic strength (Chapter 18).

9.4 Ionic bonding

If adjacent atoms are of different species, say A and B, then one or more levels on B may be lower in energy than the corresponding ones on A. Therefore, the B atoms will acquire more electrons than the A atoms, and a charge difference between the two types develops. The B atoms become negative ions, and the A atoms become positive ions. The B atoms then repulse other B atoms, the A atoms repulse A atoms, while the B atoms attract A atoms, and vice versa. For three-dimensional arrays of A and B as in sodium chloride crystals, the number of negatives equals the number of positives but they do not completely cancel one another, and there is a net attraction as first shown by Madelung (1909). This was the beginning of the quantitative theory of cohesion. The postulated distribution of charge in ionic crystals has been confirmed directly (Coppens, 1997).

9.5 Metallic bonding

In simple metals (those of the first three columns of the Periodic Table in which the s- and p-type valence electrons are separated in energy from the d-levels) the valence electrons are in anti-bonding (conduction) states, so they are delocalized and distributed in space quite uniformly. The positive ion cores of these metals are immersed as regular arrays within the uniform distribution of negative charge. However, the valence electrons are distributed non-uniformly in energy (momentum) because of the Pauli Principle.

X-ray scattering has determined this distribution directly (Vos and McCarthy, 1995). In simple metals, the valence electrons have wavelengths λ that vary from the macroscopic dimensions of a specimen to microscopic atomic lengths. Their kinetic energies are inversely proportional to these wavelengths, as indicated by the Heisenberg Principle (Equation (8.8)). Letting **k** be a wave vector with $|\mathbf{k}| = k = 2\pi/\lambda$, the kinetic energy (momentum) is proportional to k^2. The experimental verification of this is shown in Figure 9.6. It is this

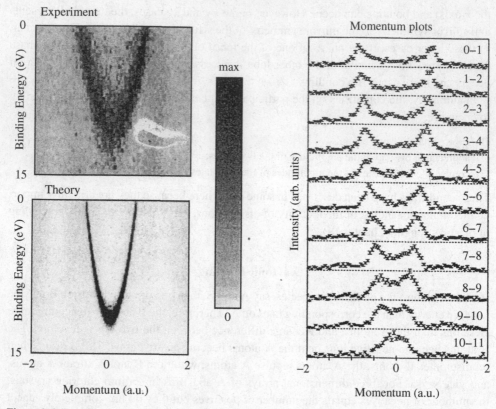

Figure 9.6 Experimental verification of the parabolic dependence of the electron energy on momentum in a nearly free-electron metal (aluminum). After Vos and McCarthy (1995). Measurements were from electron momentum spectroscopy (EMS).

energy density that keeps metals from collapsing under the Coulombic attraction between the positive ion cores and the valence electrons. This is sometimes called the "Schrödinger pressure".

9.6 London forces

When two neutral molecules (or atoms) are in close proximity they induce electric dipoles in one another, and this creates an attraction between them. In each one, the center of the negative charge fluctuates in its position relative to the positive nuclei. Thus a fluctuating electric field is emitted by each of them which interacts with the fluctuating field in the other one. When the fields are in phase, the energy of the system is lower than when the fields are out of phase, so they are attracted to each other. This leads to the formation of molecular crystals (the fourth type of bonding), to the bonding of molecular filaments into bundles, and to the bonding of two-dimensional molecular lamellae to form flakes (graphite, for example). The interactions are mediated by electromagnetic particles (photons) instead of electrostatic particles (electrons).

London forces act, for example, in polymers such as polyethylene (e.g., milk bottles). In this case the chain-like molecules are bonded along their lengths by strong covalent bonds between carbon atoms, as are the hydrogen atoms bound to the carbon atoms along the sides of the covalent carbon chains. However, the hydrogenated chains are held together by the much weaker dipole–dipole bonds.

9.7 Hydrogen bonding

A fifth type of bonding that is intermediate in strength is the special case of hydrogen bonding. This plays an important role in biological materials (protein crystals, for example), and water, and will be considered in Chapter 11.

9.8 Magnetic contributions to cohesion

The angular momenta of electrons have magnetic moments associated with them, and each electron itself has both mechanical and magnetic moments because of its self-spin. Therefore, there are magnetic interactions which make small contributions to cohesion. However, these are small compared with the electrostatic contributions.

References

Atkins, P.W. (1991). *Quanta*, p. 380. Oxford: Oxford University Press.
Bond, G.C. (2000). Relativistic phenomena in the chemistry of the platinum group metals, *Platinum Met. Rev.*, **44** (4), 146.
Brown, A.S. and Spackman, M.A. (1990). An electron-density study of germanium: evaluation of the available experimental data, *Acta Crystallogr. A*, **46**, 381.
Coppens, P. (1997). *X-Ray Charge Densities and Chemical Bonding*. New York: Oxford University Press.
Coulson, C.A. (1952). *Valence*, p. 70. Oxford: Clarendon Press.
Craig, D.P. and Thirunamachandran, T. (1998). *Molecular Quantum Electrodynamics*. Mineola, NY: Dover Publications.
Deutsch, M. (1992). Electronic charge distribution in crystalline silicon, *Phys. Rev. B*, **45**, 646.
Gray, H.B. (1973). *Chemical Bonds: An Introduction to Atomic and Molecular Structure*. Menlo Park, CA: W.A. Benjamin.
Heisenberg, W. (1930). *The Physical Principles of the Quantum Theory*, trans. C. Eckart and F.C. Hoyt. New York: Dover Publications.
Lu, Z.W. and Zunger, A. (1993). Electronic charge distribution in crystalline diamond, silicon, and germanium, *Phys. Rev. B*, **47** (15), 9385.
Madelung, E. (1909). *Gött. Nach.*, 100.
Parr, R.G. and Yang, W. (1989). *Density-Functional Theory of Atoms and Molecules*, p. 229. New York: Oxford University Press.
Spackman, M.A. (1991). The electron distribution in diamond: a comparison between experiment and theory, *Acta Crystallogr. A*, **47**, 4200.

Stewart, R.F. and Spackman, M.A. (1981). Charge density distributions. In *Structure and Bonding in Crystals*, ed. M. O'Keefe and A. Navrotsky, Volume I, Chapter 12. New York: Academic Press.

Van Arkel, A.E. (1956). *Molecules and Crystals in Inorganic Chemistry*. New York: Interscience. Also, Ketelaar, J.E. (1958). *Chemical Constitution, An Introduction to the Theory of the Chemical Bond*, 2nd edn. New York: Elsevier.

Vos, M. and McCarthy, I.E. (1995). Observing electron motion in solids, *Rev. Mod. Phys.*, **67** (3), 713. See also, *Am. J. Phys.*, **65** (6) (1997).

Weiner, J.H. (1983). *Statistical Mechanics of Elasticity*. New York: Wiley.

10

Intramolecular cohesion

The basic element of cohesion is the diatomic molecule because the same bonding principles that apply to it often apply to more complex bonding situations, particularly when only covalent bonds are present. However, when long-range forces are present new factors must be considered. In particular, for atoms with large polarizabilities, London dipole–dipole forces must be considered. Furthermore, at surfaces, plasmons (collective oscillations of electrons and positive ions) may become important.

Even for the relatively simple case of diatomics, there are no exact analytic solutions of the Schrödinger equation (except for the molecular ion, H_2^+). This has resulted in a plethora of approximate solutions. All of them are effective for interpolation, but not for extrapolation. This is because they all have ad hoc features. Only a few of the approaches can be outlined here. For others, the books by Slater (1963), Burdett (1995), Atkins (1983), Pettifor (1995), Sutton (1993), Parr and Yang (1989), Pauling and Wilson (1983), and others, can be consulted.

10.1 Covalent diatomic molecules (time independent)

10.1.1 Hydrogen molecular ion (H_2^+)

Although this is not a complete diatomic molecule, it is an important case because an exact solution of the Schrödinger equation exists for it (Slater, 1963). This is possible because it is a quasi-two-body molecule consisting of a pair of stationary nuclei with one electron swarming around them. It is the most simple example of the bonding of two protons through the exchange of a single electron. The two nuclei are so heavy compared with the electron, that the pair behaves approximately like a single particle whose length is a parameter in the determination of the minimum energy (this is called the Born–Oppenheimer approximation). In this case, the Schrödinger equation can be solved exactly (but it cannot be solved exactly for the neutral molecule, H_2, which at minimum consists of three particles).

Either of two approaches can be used for finding the binding energy of the molecular ion. One is the static approach of forming a linear combination of the amplitude functions of the two atoms. The other is the dynamic approach of considering the hopping of the binding electron back and forth between the two nuclei (Chen, 1991). The two approaches are equivalent because for both of them the electron is located, with highest probability, midway between the two nuclei.

One quasi-static approach from which we can learn quite a bit, is simply to use the Heisenberg Principle without solving the Schrödinger equation at all. This approach is Kimball's method (1950). Although it drastically simplifies the problem, it contains the important physical factors, and yields results that are surprisingly close to the experimental measurements. It is in the spirit of the matrix version of quantum mechanics. Thus it is equivalent to approximately solving Schrödinger's equation. It starts by assuming de Broglie's relationship for particles of mass m and velocity v:

$$\lambda = h/mv \tag{10.1}$$

h being Planck's constant, and λ being the wavelength of the probability amplitude for the electron. In addition to its mass, the electron has a charge q. Its spin will be neglected. The electron's total energy U is the sum of its potential energy, V, and its kinetic energy, T:

$$U = V + T \tag{10.2}$$

In this case, V is Coulombic, and since both the electrons and the protons have unit charge, it is given by (cgs units, or atomic units):

$$V = \pm q^2/r_{ij} \tag{10.3}$$

where r_{ij} is the distance between two particles of a pair, and $+$ is used if $i = j$ while $-$ is used if $i \neq j$.

From the de Broglie relation, the kinetic energy is (Chapter 7):

$$T = mv^2/2 = (h/\lambda)^2/2m \tag{10.4}$$

and for a particular coordinate, say x, the momentum is $p_x = mv_x$.

The localization of the electron along the x-coordinate can be described by means of a normal distribution function of the form: $\exp\{-[(x - x_0)/\Delta x]^2/2\}$ where Δx is the standard deviation of the distribution of x around the average, x_0. Similarly, the standard deviation of the momentum is Δp_x. Then, the Heisenberg Principle can be used to find the kinetic energy of a spherical charge distribution of nominal radius R (Chapter 7):

$$T = (3h/4\pi R)^2/2m \tag{10.5}$$

The Coulombic, or electrostatic, energy has three contributions for a cloud of electrons with protons immersed in it: first, the proton–proton repulsion; second, the electron–proton attraction; third, the electron–electron repulsion. Outside the cloud of radius, R, the electric field, E, as the cloud is approached along a radial line, is the same as if the charge in the cloud were concentrated at its center (Gauss's Theorem). Inside the sphere of radius R, the electric field decreases with distance, r, from the center in proportion to the amount of charge contained within the smaller sphere. Thus the electric field is (cgs units):

$$\begin{aligned} E &= (-q/r^2) & \text{for } r \geq R \\ &= (-qr/R^3) & \text{for } r \leq R \end{aligned} \tag{10.6}$$

The corresponding potential energies are obtained by integration:

point–point

$$V = -q/2r \tag{10.7}$$

point–cloud

$$V = -q/2r \qquad \text{for } r \geq R$$
$$= -(q/2R)[(3/2) - (1/2)(r/R)] \qquad \text{for } r \leq R \tag{10.8}$$

cloud–cloud

$$V = \int (\text{potential})_1 \times (\text{charge density})^2 \, dv = +q/2r \qquad (r \geq 2R, \text{no overlap})$$
$$= +(q/2R)[(12/5) - (r/R)^2 + (3/8)(r/R)^3 - (1/80)(r/R)^5] \qquad (r \leq 2R) \tag{10.9}$$

The cloud–cloud case is not used here, but will be used shortly for the H_2 molecule.

The simplified H_2^+ model is formed by combining the above elements. Start with a spherical cloud of negative charge of radius R. The cloud contains a total charge equal to one electronic charge, equal to one atomic unit of charge. We form a hydrogen molecular ion by embedding two protons spaced symmetrically about the center of the cloud, and a distance b apart. The first step is to find the distance, b, followed by optimizing the size of the charge cloud to give the minimum total energy of the configuration. This minus the energy of the dissociated configuration gives the dissociation energy.

In atomic units, the force of the proton–proton repulsion is $+1/b^2$, and the force of interaction between the two protons and electron cloud is $-2(dV/dr)_{\text{p-c}} = -2r/R^3 = -b/R^3$ from Equation (10.8). Equating the forces yields $b = R$, so the protons straddle the center point of the electron cloud lying at the 1/4 and 3/4 points along its diameter.

The kinetic energy of the charge cloud from Equation (10.5) is $9/4R^2$, or in terms of the bond length b it is $+9/4b^2$. The electrostatic (potential) energy of the proton–proton repulsion is $+1/b$. And, the electrostatic energy of the two protons and the electron cloud is $-(2/R)\{3/2 - 1/2(b/2R)^2\} = -11/4b$, so the sum of the two electrostatic energies is $7/4b$. Setting the derivative of the total energy equal to zero gives an equilibrium bond length of 18/7 atomic units, or 1.36 Å which is 22% larger than the experimental bond length of 1.06 Å. Substituting back into the expression for the total energy gives $U_T = -0.34$, so the dissociation energy is $-0.29 = -3.9$ eV which is 31% higher than the experimental value of 2.7 eV. Considering the simplicity of this model, its agreement with experiment is good, as good as some approximate solutions of the Schrödinger equation.

10.1.2 Hydrogen molecule (H₂)

Applying the same method to the hydrogen molecule yields similar results. The start is the same, but with two electrons. The model molecule is made by centering a charge cloud of radius R, containing two electrons, at the coordinate origin, and putting one proton at $+r$, and the other at $-r$. To find r in terms of R, the equilibrium of forces is considered as in the previous case. Atomic units are used: the energy unit is $(2\pi^2 m q^4)/(h)^2$; the length unit is $(h/q)^2(4\pi^2 m)$.

The repulsive forces (proton–proton plus electron–electron) are:

$$2/(2r)^2 = 1/(2r^2)$$

while the attractive proton–electron forces are:

$$2(\text{grad } V_{\text{point–cloud}}) = 2(2r/R^3)$$

thus, at equilibrium, $r = R/2$ which is used to obtain an expression for the total energy of the model molecule. This total energy is the sum of the kinetic, proton–proton, proton–electron, and electron–electron energies:

$$U = 2[9/(4R^2)] + 2(1/R) - 4\{[3R^2 - (R/2)^2]/R^3\} + 12/5R \qquad (10.10)$$

To find the value of R at which this is minimized, $dU/dR = 0$ is solved, yielding $R^* = 15/11$. This is substituted back into the expression for U, to obtain $U^* = -121/50$, and a dissociation energy of $21/50$. Comparing the model values with measured ones,

	Model	Measured
Proton separation (Å)	0.72	0.74
Total energy (eV)	33.0	32.0
Dissociation energy (eV)	5.7	4.7

we see that this simple model gives remarkably good results. However, the reason for presenting it is not that it is accurate but that it shows the nature of the energy terms that are involved, and their relative magnitudes.

It is of interest to calculate the bonding energy density of the hydrogen molecule. The volume is about $1.7 \, \text{Å}^3$, so the dissociation energy density is about $2.8 \, \text{eV/Å}^3$. This corresponds to a bonding energy density (internal pressure) of about 2.8 Mbar.

10.2 Time dependent theory (resonance)

For a more general view of atomic cohesion, it is useful to consider the fact that the positive nucleus inside an atom is not rigidly positioned at the center of the electron swarm, nor are the valence electrons rigidly attached to particular atoms (nuclei). Therefore, the nuclei can oscillate inside of atoms and if their oscillation frequencies match those of nearby atoms, resonances can occur. Also, if the energy levels for the electrons match between two (or more) atoms, then electrons can tunnel back and forth resonantly from one atom to the other in an identical pair. These two kinds of resonance might be called "weak" and "strong". We consider weak resonances first.

10.2.1 Weak resonance (London forces)

In the presence of an electric field, E_x, directed in the $+x$ direction, the nucleus of an atom will tend to move in the same direction, while the centroid of the electron swarm will tend to move in the $-x$ direction, so the atom becomes polarized. Its shape changes slightly (typically from spherical to ellipsoidal) and it acquires a dipole moment, $\mu_x = qx$, where x is the separation distance, and q is the charge of the electron. If the electric field is removed, the dipole moment vanishes. Thus the system behaves like a spring. However, the response to the applied electric field is not instantaneous since the electron has an inertial mass. The combination of spring and mass constitutes a harmonic oscillator, or "pendulum", with angular frequency, $\omega_a = \sqrt{(k/m)}$, where k is the spring constant for the restoring force and m is the electronic mass. Also, the induced dipole moment is proportional to the applied field: $\mu_x = \alpha E_x$, where the coefficient α is called the polarizability. In general, the polarizability is a tensor, α_{ij}.

If two atomic oscillators are near one another, they will interact because each oscillator will emit electromagnetic waves that will travel to the other one and either reinforce its oscillations, or interfere with them. If they interact in phase (reinforcement) they will attract each other, whereas if the interaction is out of phase they will repulse each other. This is analogous with the behavior of two mechanical pendulums that are coupled together by a "weak spring" such as a common mounting bracket that is elastic.

As each atom vibrates it emits an electromagnetic wave (photon) which travels to and envelopes the other atom of the pair. The other atom responds by becoming polarized with a dipole moment $\pm\mu$. Let us label the atoms A and B. Then when the moment on A is $+\mu$, and the moment on B is $-\mu$, the mutual vibrations are out of phase (anti-symmetric), and the atoms attract one another. When the moment on A is $+\mu$, while that on B is $+\mu$, the vibrations are in phase (symmetric), and the atoms repulse one another.

Now let us introduce a pinch of quantum mechanics to the system. The resonant frequencies of the atoms are $\omega = \sqrt{(k/m)}$. If they are taken to be quantized harmonic oscillators, their energy levels will be spaced at intervals of $\hbar\omega$ where $\hbar = h/2\pi$, and the set of energy levels is $(n + 1/2)\hbar\omega$ $(n = 0, 1, 2, \dots)$. The lowest energy state is now $n = 0$, with energy $\hbar\omega/2$ (the zero-point energy), and transitions between the levels occur in amounts $\hbar\omega$ with $n = \pm 1$. Thus the atoms emit and absorb photons of energy, $\hbar\omega$, which they can be said to "exchange", and they are attracted to one another (or repulsed) by exchange forces. These forces are relatively weak, so this is a "weak interaction". If the atoms exchange electrons, rather than photons, the interaction becomes "strong". This will be discussed below.

In Chapter 11 it is shown that the interaction energy between two oscillating dipoles is given by:

$$U_{\text{dip-dip}} \sim -\alpha^2/r^6 \tag{10.11}$$

so it depends strongly on their polarizabilities, and very strongly on the distance between them. If the distance increases by a factor of two, the interaction energy declines by a factor of 64.

Elastic stiffness

Also in Chapter 11 an approximate expression for the polarizability is derived in the context of quantum mechanics using perturbation theory. The result is:

$$\alpha = (2/m)(qh/2\pi\Delta)^2 \tag{10.12}$$

where m and q are the mass and charge of the electron, respectively, h is Planck's constant, and Δ is the difference in energy between the ground state and the first excited state. The inverse square dependence on Δ means that the polarizability is quite sensitive to the excitation energy. This is consistent with the fact that substances with high polarizabilities are often colored, or black.

10.2.2 Strong resonance (hopping)

Turning to the strong interaction case, the theory of molecular orbitals will be introduced. Notice that the word "orbitals" is used, not the word "orbits". This is because the electrons are not considered to be moving in tracks around the nuclei. They might be viewed as swarming within volumetric regions that are defined by the probability amplitudes which are solutions of Schrödinger's equation. In the time domain, the amplitudes, ψ, change according to first-order kinetics. That is, the rate of change $d\psi/dt$ is proportional to ψ itself with the rate constant determined by the Hamiltonian (energy) operator.

10.2.3 Hydrogen molecular ion (H_2^+)

For this most simple of molecules, let the distance between its two protons be b. Label the protons A and B. Let the probability amplitude when the electron is primarily associated with A be ψ_A, and when it is primarily associated with B be ψ_B. Each of these is normalized so $\int \psi_x \psi_x^* \, dv = 1$ where $x = $ A or B, and the ψ_x are orthogonal so $\int \psi_A \psi_B^* \, dv = 0$. The energy of each electron, E_x, when the atoms are widely separated is given by $H\psi_x = E_x \psi_x$ where H is the Hamiltonian operator. This relation states that when H operates on ψ_x, the result is the total energy E_x (a number), times the probability amplitude.

Since it represents a wave, the probability amplitude depends on both position and time. It can be written $\psi = \psi(x, y, z, t) = \psi_0 e^{-i\omega t}$ where ω is the energy divided by Planck's constant, ω is the wave's frequency, $E/\hbar$, and $\hbar = h/2\pi$. Remember that $e^{-i\theta} = \cos\theta - i\sin\theta$. To obtain the time dependence of ψ we differentiate it with respect to time:

$$d\psi/dt = -i\omega\psi_0 e^{-i\omega t} = -(i E/\hbar)\psi = -(H/\hbar)\psi \tag{10.13}$$

To describe the molecular combination of two atoms plus one electron we form a new amplitude function, $\Psi(r)$, called a molecular orbital. It adds the amplitudes of the individual atoms, forming a linear combination of atomic orbitals (LCAO):

$$\Psi = \alpha\psi_A + \beta\psi_B \tag{10.14}$$

Like the ψ, it is a solution of the Schrödinger equation. Here, α and β are coefficients that describe the relative amounts of the amplitudes, and the ψ_x are declining exponentially with the distances r from each of the positive protons. Thus each amplitude has the form: $\psi(r) = (1/\sqrt{\pi})e^{-r}$ (atomic units, 1s state with $n = 1$, $l = 0$, $m = 0$). This form for the amplitude has been confirmed by direct measurements of the charge density (Coppens, 1997) (Figure 6.6).

As the two protons are brought together from a large separation distance, until they "touch" at about $r = 2\pi a_0$ (where a_0 is the Bohr radius, equal to one atomic unit of distance), the single electron of the molecular ion binds the protons by hopping back and forth from the vicinity of proton A to that of proton B. There is a potential energy barrier resisting the hopping, but the electron can tunnel through the barrier. The rate of hopping depends on the height and breadth of the barrier, as well as the attempt frequency, so it is relatively slow when the separation distance is large, but it speeds up as the protons move closer together, finally reaching about 10^{16} s^{-1}. The frequency can be calculated from the bonding energy. If U_b is the bonding energy, then dividing it by Planck's constant h gives the approximate maximum hopping (tunneling) frequency (Chen, 1991).

Since the hydrogen molecular ion is symmetric, the coefficients in the amplitude equation (10.14) are equal, say $\alpha = \beta = 1$, so the probability density for the molecular system is obtained by squaring the amplitude (remembering that the ψ are real functions):

$$\Psi^2 = \psi_A^2 + \psi_B^2 + \psi_A\psi_B \qquad (10.15)$$

In this equation the mixed term, $\psi_A\psi_B$, represents the average probability distribution in the space between A and B as the electron hops (tunnels) back and forth. In order for this term to be appreciable the two amplitudes must overlap. They do this along the line passing through the two protons, so electron amplitude builds up between the atoms. This provides part of the bonding by attracting the positive protons until they become close enough together that their mutual repulsion balances the attraction. However, most of the bonding comes from contraction of the atomic amplitudes as they mutually interact. The contractions increase the magnitudes of the electrostatic interactions of the electrons and the protons (negative energies). This viewpoint was originally proposed by Heisenberg in 1926. A detailed treatment of the theory may be found in the book by Pauling and Wilson (1983), and a more modern treatment in connection with tunneling microscopy in a paper by Chen (1991). An outline of the theory is given below.

The probability amplitude for finding the electron on A is $\psi_A = (1/\sqrt{2})\exp(-r_A)$, and the same for B with the A replaced by B. For the system of three particles, an amplitude function can be formed simply by adding together appropriate fractions of the individual amplitudes. Thus, if $\alpha = 1$, while $\beta = 0$, the electron is localized at A. If $\alpha = \beta = 1/\sqrt{2}$, then the electron is 50% at each A and B (to satisfy the normalization condition, $\alpha^2 + \beta^2 = 1$).

In the molecular ion, the electron must be distributed, on average, over the whole system. So it must "hop" back and forth between A and B. Thus, α and β must vary in time. Call ω_h the hopping frequency. The associated energy is $\hbar\omega_h$. However, there are two hopping directions: $\alpha \to \beta$, and $\beta \to \alpha$, plus hopping in place, $\alpha \to \alpha$, and $\beta \to \beta$. The latter represent stationary states. Therefore, the states on A and B are connected by a matrix of hopping frequencies:

$$\begin{bmatrix} \omega_{\alpha\alpha} & \omega_{\alpha\beta} \\ \omega_{\beta\alpha} & \omega_{\beta\beta} \end{bmatrix}$$

where $\omega_{\alpha\alpha} = \omega_{\beta\beta}$, and $\omega_{\alpha\beta} = \omega_{\beta\alpha}$.

Since the kinetic Schrödinger equation for the amplitude is first order, the coefficients of the amplitudes follow the first-order kinetics given by the coupled differential equations:

$$d\alpha/dt = -i(\omega_{\alpha\alpha}\alpha + \omega_{\alpha\alpha}\beta)$$
$$d\beta/dt = -i(\omega_{\alpha\beta}\alpha + \omega_{\alpha\alpha}\beta) \tag{10.16}$$

That is, the rates of change of the coefficients are proportional to the current values of each coefficient; and the rate constants are the hopping frequencies. Note that because of the symmetry of the configuration there are only two independent hopping frequencies. This pair of equations can be solved by first forming their sum and difference. This yields two homogeneous equations that are readily integrated. Recalling Euler's relation: $e^{ix} = \cos x + i \sin x$, and setting $t = 0$, the expressions for the coefficients are:

$$\alpha = [\exp(-i\omega_{\alpha\alpha}t)] \cos(\omega_{\alpha\beta}t)$$
$$\beta = [-i \exp(-i\omega_{\alpha\alpha}t)] \sin(\omega_{\alpha\beta}t) \tag{10.17}$$

Since the probability, P_A, of finding the electron in state A is $\alpha\alpha^{\star} = |\alpha|^2$, and $P_B = \beta\beta^{\star}$ (where $x^{\star}$ is the complex conjugate of x), the time-dependent probabilities are:

$$P_A = \cos^2(\omega_{\alpha\beta}t)$$
$$P_B = \sin^2(\omega_{\alpha\beta}t) \tag{10.18}$$

These equations indicate that initially ($t = 0$) the charge density at proton A is unity, and zero at proton B. Then the electron begins to tunnel (hop) to proton B, so the fractional charge at A decreases with time, and increases at B, until it becomes unity at B and zero at A. This state is reached at $t = 1/4\omega_{\alpha\beta}$. The trend then reverses until the initial state is restored at $t = 1/2\omega_{\alpha\beta}$, whereupon the cycle repeats.

The driving force for the hopping is the potential of the additional proton in the molecular ion. This additional potential depends on the distance, b, between the two protons which is the bond length. When $b \gg 1$ (atomic units), the rate of hopping between A and B is very small so the system consists essentially of a hydrogen atom and a distant proton. The atom–proton attraction is correspondingly small.

In order for the electron to hop from A to B (or inversely), since it is tightly bound to A and its zero-point energy is small, it must tunnel through the potential barrier. The tunneling probability is a strong function of the distance, b. For small b, it varies as b^{-4}, and for larger b as $\exp(-2b)$. For a full discussion of the tunneling probability, see Chen (1991).

10.2.4 Hydrogen molecule (H_2)

The addition of one more electron to the system of one electron and two protons complicates the problem so that an exact solution is not possible. There are too many interactions and they are interdependent. Since the electrons have the same charge signs they repulse one another, so both of them are associated with the same atom only a relatively small fraction of the time. Therefore, it becomes appropriate to call the hopping from one atom to the other "exchange".

10.2.5 Morse potential

The most generally successful function that is used to describe the variation of the energy with distance is a combination of exponentials called the Morse potential (1929). It is instructive to start with two factors that make up this function. The factors are exponentials with the form:

$$f = \exp[-\beta(r - b)] \tag{10.19}$$

where b is the equilibrium bond length, r is the nuclear separation distance, and β is a constant (reciprocal relaxation distance). Then the attractive part of the Morse potential is $A = 2f$, and the repulsive part is $R = f^2$ (see Figure 10.1). If the energy needed to separate the two atoms of a diatomic molecule (the dissociation energy) is called D, then the bonding

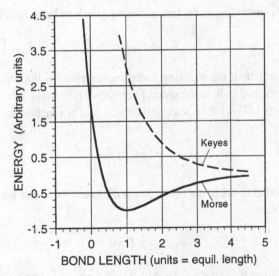

Figure 10.1 Illustration of the Morse bonding potential plotted together with the Keyes anti-bonding potential. Note that the thermal dissociation energy here is one energy unit, and the single photon dissociation energy is three energy units. In the latter case, the steep slope of the Keyes curve at unit bond length indicates a very high dissociation (Schrödinger) pressure.

form of the Morse potential function, $E_b(r)$ takes the simple form:

$$E_b(r) = D(f^2 - 2f) \tag{10.20}$$

and Keyes (1958) has shown that an approximate anti-bonding form is given simply by changing the sign:

$$E_a(r) = D(f^2 + 2f) \tag{10.21}$$

When $r = b$, $f = 1$, so $E_b = -D$, and $E_a = +3D$. Thus, the splitting of the atomic energy levels is very asymmetric when a molecule is formed. That is, the anti-bonding orbital is destabilized three times the amount that the bonding orbital is stabilized.

It is of interest to explore the strength of the Morse type bond since its form is approximately universal (Ferrante, Smith, and Rose, 1983). By strength is meant the force or stress needed to break the bond. The force, F, is the slope of the energy curve:

$$\begin{aligned} F(r) = dE_b/dr &= (dE_b/df)(df/dr) \\ &= [2D(f-1)](-\beta f) = -2\beta Df(f-1) \end{aligned} \tag{10.22}$$

and the second derivative, evaluated at $r = b$, is the force constant which determines the vibrational frequency. The second derivative is:

$$d^2 E_b/dr^2 = -2D\beta^2 f(2f - 1) \tag{10.23}$$

but at $r = b$, $f = 1$, so the force constant at the equilibrium bond length is:

$$k_b = -2D\beta^2 \tag{10.24}$$

As the bond is stretched, the resisting force increases until the inflection point of the potential energy is reached (where the second derivative is zero) at $f = 1/2$ and $r_{inflection} = b + (\ln 2)/\beta = b + 0.347\,(\text{Å})$. The maximum force that is reached at the inflection point is:

$$F_{max} = -\beta D \tag{10.25}$$

This is the force needed to break the bond. Since the cross-sectional area is approximately $4b^2$, the breaking stress is:

$$\sigma_{max} \approx \beta D/4b^2 \tag{10.26}$$

For the case of H_2, taking $\beta = (\text{Bohr radius})^{-1} = 1.89\,\text{Å}^{-1}$, $b = 0.74\,\text{Å}$, and $D = 4.72\,\text{eV}$, the breaking stress is 6.5 Mbar. This is a very high strength but, unfortunately, hydrogen does not form stable chains, so it has little practical significance.

10.3 Polyatomic molecules

10.3.1 Hybridization

A simple model of the bonding of hydrogen is possible because only two electrons are involved, the orbitals they occupy are spherically symmetric, and their spins can form up/down pairs so the Pauli Principle is easily satisfied. For molecules with more than two electrons the electronic structure becomes considerably more complex, and the number of structural types that can be assembled becomes very large. Of special interest for strong solids are the open structures (diamond, zinc blende, and wurtzite) that can be formed from atoms like carbon because they can form bonds with tetrahedral symmetry. The four bonds per atom in this case is the smallest number from which a uniform three-dimensional framework can be formed. Also, four is the minimum number of struts required to fix the position of a nodal point in three dimensions. Finally, carbon is the first atom in the Periodic Table that has enough electrons to form four bonds. Therefore, it has the smallest atomic volume, and the largest valence electron concentration. It is this high electron density that allows carbon to be the basis of organic chemistry, and for diamond to be the hardest known substance.

Rationalization of the tetrahedral bond configuration of carbon as diamond, along with its trigonal bond configuration as graphite, is one of the many triumphs of quantum mechanics. As an illustration of how wave behavior is intricately connected with symmetry factors it is also central to an understanding of the mechanical properties of solids. Therefore, it is described in some detail in the next several paragraphs.

The amplitude functions that form the lowest energy s-states of atoms have spherical symmetry, and those that form the next lowest energy p-states are "dumbbells" with cylindrical symmetry, each lying along one of three orthogonal axes. Thus neither of these distributions individually has tetrahedral symmetry, but if the functions that describe them are combined appropriately as linear combinations, a function with tetrahedral symmetry emerges. This was first shown in the early 1930s by Pauling (1931), and by Slater, nearly simultaneously.

The relative magnitudes of the angular wave functions that satisfy the Schrödinger equation (spherical coordinates r, θ, ϕ) are as follows:

$$s = 1$$
$$p_x = 3^{1/2} \sin \theta \cos \phi$$
$$p_y = 3^{1/2} \sin \theta \sin \phi$$
$$p_z = 3^{1/2} \cos \theta$$

A linear combination of these is also a solution, so it can be used to form a hybrid bond function, ψ_h:

$$\psi_h = a_h s + b_h p_x + c_h p_y + d_h p_z \qquad h = 1, 2, 3, 4$$

where the a_h, b_h, c_h, d_h are coefficients.

Since the Schrödinger equation is separable into radial and angular parts, the radial form of the solution is not affected by the formation of this linear combination, so it will not be written explicitly.

The hybrid bond amplitude function must be normalized; so, $\int \psi_h^2 \, dv = 1$, or:

$$a_h^2 + b_h^2 + c_h^2 + d_h^2 = 1$$

and the hybrids must be orthogonal (independent), so:

$$a_h a_k + b_h b_k + c_h c_k + d_h d_k = 0 \qquad h, k = 1, 2, 3, 4 \; (h \neq k)$$

Let the first bond hybrid ($h = 1$) lie along the x-axis, where $p_y = p_z = 0$. Then

$$\psi_1 = a_1 s + b_1 p_x$$

but, from the normalization condition:

$$(b_1)^2 = 1 - (a_1)^2$$

so the amplitude of the first hybrid bond is:

$$\psi_1 = a_1 + \left[3\left(1 - a_1^2\right)\right]^{1/2} \tag{10.27}$$

The constants can be evaluated by setting the derivative of this equal to zero, and solving for the maximum:

$$a_1 = 1/2 \qquad \text{and} \qquad b_1 = (3^{1/2})/2$$

Using these values, plus $s = 1$, and $p_x = 3^{1/2}$ along the x-axis, gives the maximum value $\psi_1 = 2$. This is 15% greater than the maximum value of p_x along the x-axis which is $3^{1/2} = 1.732$. Thus the hybrid has lower energy than either the s or the p_x orbital alone.

To find the optimum position of the next hybrid orbital, it is assumed that its maximum direction lies somewhere in the xz-plane. Then p_y makes no contribution to it, and the amplitude function takes the form:

$$\psi_2 = a_2 s + b_2 p_x + d_2 p_z$$

This lies in the xz-plane, so the angle ϕ must be $0°$ or $180°$, but the former has already been used for ψ_1, so $\phi = 180°$ in this case, and $\cos \phi = -1$:

$$p_x = -3^{1/2} \sin \theta$$

As before, $s = 1$ and $p_z = 3^{1/2} \cos \theta$. The normalization and orthogonalization conditions are respectively given by:

$$(a_2)^2 + (b_2)^2 + (d_2)^2 = 1$$

and:

$$a_1a_2 + b_1b_2 = 0$$

since $d_1 = 0$. Using the values already obtained for a_1 and b_1:

$$a_2 = -3^{1/2}b_2 \quad \text{and} \quad d_2 = \left[1 - (2b_2)^2\right]^{1/2}$$

Therefore, the second hybrid orbital is:

$$\psi_2 = -3^{1/2}b_2(1 + \sin\theta) + \left\{3\left[1 - 4(b_2)^2\right]\right\}^{1/2}\cos\theta \tag{10.28}$$

To find the b_2 and θ that make this a maximum, it is partially differentiated with respect to each of them. The two derivatives are set equal to zero giving two simultaneous equations that can be solved to obtain:

$$b_2 = -(1/12)^{1/2} \quad \text{and} \quad \sin\theta = 1/3$$

Thus $\theta = 19°28'$, and the maximum value of ψ_2 is 2.0 which is the same as for ψ_1. Also, the angle between this orbital and the first one is $(\theta + 90°) = 109°\,28'$ which is the tetrahedral angle.

Similar arguments yield ψ_3 and ψ_4 which also have maximum values of 2.0, and lie at tetrahedral angles with respect to the others.

Although this original hybridization theory of Pauling (1931) is an oversimplification, it shows that within the rules of quantum mechanics, hybridization can lead to directed bonds with tetrahedral symmetry, and therefore to stable open structures of high symmetry such as the diamond and zinc blende structures. The hybrid just discussed is called sp^3 because it can accommodate the four valence electrons of carbon, one of the s-type and three of the p-type.

Many other hybrids are possible, and some play important roles in the formation of structures with other symmetries. For example, in benzene and graphite, sp hybrid orbitals with trigonal symmetry are important. sp^2d hybrids yield planar square symmetry, and sp^3d^2 hybrids have octahedral symmetry (i.e., six-fold coordination). The latter play an important role in coordination chemistry.

Experimental evidence for the existence of these hybrids is provided by stereochemistry, and more directly by X-ray scattering (Figure 6.6). Much of the experimental evidence is summarized by Coppens (1997).

10.3.2 Chains

Linear strings of atoms are the basic structures of polymers. They also represent one-dimensional solids, and their electronic structures indicate how bands of states form as the energy states of the individual atoms of a chain split into states of slightly different energies. For chains of finite length, surface states (Tamm states) form at each end. These sometimes play an important role in fracture, and in surface effects during plastic flow.

10.3.3 Diatomic chains

We start with a very short chain, just two atoms spaced one bond length b apart. Instead of the hopping amplitude for an electron consider time independent states, ϕ_i. Since we will soon switch to a long chain, use numbers to label the atoms: 1 and 2. Thus, ϕ_1 is the probability amplitude for atom 1 alone, and ϕ_2 for atom 2 alone. Then the amplitude for the system of two atoms may be formed by either adding, or subtracting, the amplitudes of the individual atoms:

$$\Phi = \alpha_1 \phi_1 \pm \alpha_2 \phi_2 \tag{10.29}$$

where Φ is normalized so $\alpha_1^2 + \alpha_2^2 = 1$, and in the simplest case the atoms are identical so $\alpha_1 = \alpha_2$. In general the amplitudes are complex numbers, so the expression for the system probability is:

$$\Phi\Phi^\star = \alpha^2 \left(\phi_1^2 \pm 2\phi_1\phi_2 + \phi_2^2 \right) \tag{10.30}$$

and the importance of the fact that the probability is the product rather than the sum of the amplitudes becomes evident. We have the probabilities (charge densities) for the individual atoms, and in addition an overlap (interaction) term, $\pm 2\phi_1\phi_2$, having two possible values depending on whether the amplitudes interact constructively or destructively, that is, depending on the relative phases of the functions. In the constructive case, the amplitude is increased in the zone of interaction, and conversely for the destructive case. The two cases yield different energies for the system, of course.

Φ, the molecular orbital, is the probability amplitude for a particular electron orbital whose size and shape is determined by applicable quantum numbers. Just as atomic amplitudes can be either real or complex, so can molecular ones. Also, the square of the modulus, $\Phi\Phi^\star = |\Phi|^2$ is the probability of finding the electron in a small volume element, dv, providing the probability is normalized so $\int \Phi^2 \, dv = 1$. For the normalized case, $\Phi\Phi^\star$ is also the charge density in atomic units.

Each Φ has a definite energy value based on its wavelength (frequency), and given by Planck's relation. This is the energy needed to remove the electron from the molecule by ionization. The total energy of a molecule is the sum of the energies of the occupied molecular orbitals, plus the energies of interaction between the electrons (depending on the circumstances, the nuclear–nuclear interactions, and the energies of the atomic core electrons may also be added in). The electron spins play a role by limiting the number of electrons that can occupy a bonding molecular orbital. The limit is two, and their spins must be opposed.

To find the system energies, the time independent Schrödinger equation is used:

$$H\Phi = E\Phi \tag{10.31}$$

i.e., applying the energy operator H to the amplitude function gives the total energy times the amplitude (e.g., differentiating an exponential function yields a constant times the function).

Multiplying both sides of Equation (10.31) by Φ, and integrating both sides over the volume, dv, then solving for the energy gives:

$$E = \int \Phi H \Phi \, dv \Big/ \int \Phi^2 \, dv \qquad (10.32)$$

and, if Φ is known (as well as the energy operator H), the energy is determined.

Although Φ is not known exactly for any molecule except H_2^+, the Variational Principle helps to limit errors. Suppose that E_o is the minimum energy of the molecular system. Then, let an amplitude function, Φ_i, be guessed, and its energy E_i calculated from Equation (10.32). By the Variation Principle, E_i is always greater than or equal to E_o (never less than):

$$E_i \geq E_o$$

Therefore, if E_o is known from experiment, by systematically varying Φ_i to find the function that gives minimum energy, the true value for Φ can be approximated more and more closely.

Putting Equation (10.29) into Equation (10.32), for the homodiatomic case where $\alpha_1 = \alpha_2 = \alpha$ (the α cancel out):

$$E = \left(\int \phi_1 H \phi_1 \, dv \pm 2 \int \phi_1 H \phi_2 \, dv + \int \phi_2 H \phi_2 \, dv \right) \Big/$$

$$\left(\int \phi_1^2 \, dv \pm 2 \int \phi_1 \phi_2 \, dv + \int \phi_2^2 \, dv \right)$$

Since the atomic amplitudes are normalized, $\int \phi_1^2 \, dv = \int \phi_2^2 \, dv = 1$. Also, if the overlap integral is called S:

$$S = \int \phi_1 \phi_2 \, dv$$

then the normalization factor for Φ (the denominator) becomes $2(1 \pm S)$. Also, if the individual energy terms in the numerator are designated $H_{11} = H_{22}$, and H_{12}, then the numerator becomes $2(H_{11} \pm H_{12})$, and the expression for the energy is:

$$E = (H_{11} \pm H_{12})/(1 \pm S) \qquad (10.33)$$

Here $|S| < 1$, except when ϕ_1 and ϕ_2 coincide, then $S = 1$. Thus the S_{ij} integrals $(0 \leq S_{ij} \leq 1)$ measure the overlap of the amplitude functions of the two atoms.

The H_{ij} are called the energy matrix elements where the H_{11} and H_{22} refer to the energies of the individual atoms. They are assumed to change relatively little as the atoms come together in the molecule. H_{12} refers to the interaction of the atomic orbitals, and is called the resonance, or hopping, integral.

Equation (10.33) indicates that there are two values for E: a bonding value E_b when the plus signs are used (since the H_{ij} are negative while S is positive), and an anti-bonding value E_a when the minus signs are used. Since the bonding energy is negative, strong bonds are favored by S being close to unity, that is, by large overlaps.

For accurate calculations, it should not be assumed that H_{11} and H_{22} do not change as a molecule forms. The amplitudes at the individual atoms tend to shrink as the distance between the atoms decreases, thereby lowering the electrostatic energy (increasing the bond strength).

The bond length is determined by the forms of the ϕ. In order for the probabilities, ϕ^2, to be normalizable, the tails of the amplitude functions must approach zero asymptotically. The most simple functions that do this, and which satisfy the wave equation, are exponentials.

10.3.4 Polyatomic chains

Consider a hydrogen polymer consisting of a row of N hydrogen atoms, numbered 1, 2, 3,..., $N - 2$, $N - 1$, N, and spaced evenly a distance b apart (Figure 10.2). Note that this is an artificial model since such a row would spontaneously dimerize (i.e., form pairs, H_2 molecules). That is, it would undergo what is called the Peierls instability. Nevertheless, the model illustrates what happens when one passes from a short to a long chain (Feynman, Leighton, and Sands, 1963).

An equation for the molecular orbital, Φ, of the chain in terms of atomic orbitals, ϕ (analogous with Equation (10.29)) is:

$$\Phi = c_1\phi_1 + c_2\phi_2 + c_3\phi_3 + \cdots = \sum_{j=1}^{N} c_j\phi_j \tag{10.34}$$

where the ϕ_j are spherical, s-type amplitude functions. The coefficients c_j need to be determined, as well as the energy of the moleculer state, Φ.

To determine the c_j and find the energy we need to solve the Schrödinger equation:

$$H\Phi = E\Phi \tag{10.35}$$

Substituting Equation (10.34) into Equation (10.35) yields:

$$\sum_{j=1}^{N} c_j H\phi_j = E \sum_{j=1}^{N} c_j\phi_j \tag{10.36}$$

For the dimer, $N = 2$ (the hydrogen molecule), this becomes:

$$H(c_1\phi_1 + c_2\phi_2) = E(c_1\phi_1 + c_2\phi_2)$$

Since the energy at a particular site j depends on the charge density there, and this is proportional to $|\phi^2|$, we can evaluate Equation (10.36) by multiplying both sides by the amplitude for a particular site, say ϕ_p, and then integrating both sides (where H is again

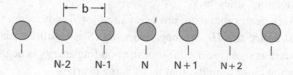

Figure 10.2 Linear chain of atoms with spacing b.

the Hamiltonian (energy) operator, E is the energy, and the c_j are constants):

$$\sum_j^N \int c_j(\phi_p H \phi_j)\, dv = E \sum_j^N \int c_j(\phi_p \phi_j)\, dv \qquad (10.37)$$

In doing this we remember that the ϕ are both normalized and orthogonal. Therefore, for the overlap integrals we have: $\int \phi_p \phi_p\, dv = 1$ and $\int \phi_p \phi_j\, dv = 0$. Hence, as j sweeps through the various sites of the right-hand side in Equation (10.37) all of them drop out except the pth one. For example, for $p = 1$, the right-hand side becomes Ec_1. However, if the atoms begin to overlap too much, the overlap integrals, $\int \phi_p \phi_p\, dv$, no longer equal unity, and the analysis becomes more complex.

On the left-hand side, there are two kinds of matrix elements:

atomic (or on-site) $\qquad H_{pp} = \displaystyle\int \phi_p \phi_p^*\, dv = \epsilon$

atomic (hopping) $\qquad H_{jp} = H_{pj} = \displaystyle\int \phi_j \phi_p\, dv = \eta$

Here it is assumed that hopping only occurs between nearest neighbors. The hopping integrals for longer distances are small, taken here to be zero. So, for various positions, $p = k = 1, 2, 3, \ldots, N$:

$$
\begin{aligned}
p &= 1 & \epsilon c_1 + \eta c_2 &= E c_1 \\
p &= 2 & \eta c_1 + \epsilon c_2 + \eta c_3 &= E c_2 \\
p &= 3 & \eta c_2 + \epsilon c_3 + \eta c_4 &= E c_3 \\
&\ \vdots \\
p &= j & \eta c_{j-1} + \epsilon c_j + \eta c_{j+1} &= E c_j \\
&\ \vdots \\
p &= N-1 & \eta c_{N-2} + \epsilon c_{N-1} + \eta c_N &= E c_{N-1} \\
p &= N & \epsilon c_{N-1} + \eta c_N &= E c_N
\end{aligned}
$$

Except for the top and bottom ones, the rest of these equations have the form of the equation for $p = j$, that is:

$$c_{j-1} - z c_j + c_{j+1} = 0 \qquad (10.38)$$

where $z = (E - \epsilon)/\eta$.

The special equations at the top and bottom ends of the set can be eliminated by combining the ends so the chain becomes an endless loop of N atoms. The loop is then said to have a periodic boundary condition because a trip around it comes back to the same place (Figure 10.3). The Nth position becomes the same as the zeroth position, so:

$$c_N = c_0 \qquad \text{and} \qquad c_{N+1} = c_1$$

From one atom of the ring to the next the electron density varies periodically, and the amplitude varies periodically, so the c_j must vary periodically. Thus the solutions of Equation (10.37) can be expected to be exponential functions (or, according to Euler's

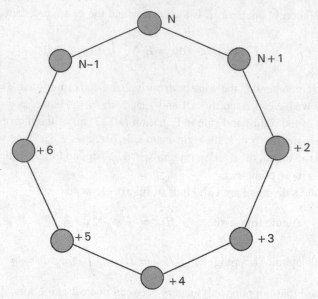

Figure 10.3 Loop of N atoms with periodic boundary condition since the Nth atom occupies both the zeroth and the eighth positions so the configuration at each atom is the same as for every other atom.

equation, sines and cosines). Let the position of the jth atom be x_j, and its nearest neighbors $x_j - b$ and $x_j + b$. The bond length is b. Also, let the wavelength of the probability amplitude be λ, and its wave number $k = 2\pi/\lambda$. Then solutions of Equation (10.38) have the form: $c(x_j) = \exp(ikx_j), c(x_j + b) = \exp[ik(x_j + b)]$, etc. Substituting these into Equation (10.38) gives:

$$\exp[ik(x_j - b)] - z\exp(ikx_j) + \exp[ik(x_j + b)] = 0$$

Dividing out the $\exp(ikx_j)$, and using Euler's equation:

$$z = -(e^{-ikb} + e^{ikb}) = -2\cos(kb) = (E - \epsilon)/\eta$$

so

$$E = \epsilon - 2\eta\cos(kb) \tag{10.39}$$

This indicates that the energy varies from $(\epsilon - 2\eta)$ to $(\epsilon + 2\eta)$ as k goes from zero to $\pm\pi/b$. However, the number of atoms in the loop, N, cannot be less than two. For this case, $k = 0$ or $\pm\pi/b$, and there are two values of E: $\epsilon \pm 2\eta$. Since η is negative, $\epsilon + 2\eta$ is a bound state, while $\epsilon - 2\eta$ is anti-bound. As N increases, for each new value of N, one new value of energy appears, alternately higher or lower than ϵ. Thus, for large N, the band of energy $\epsilon \pm 2\eta$ becomes filled with levels with only small increments between the levels (Figure 10.4). Half the band is made up of bonding levels, and half of anti-bonding levels. This is the principal result of extending the simple theory from a diatomic molecule ($N = 2$) to a polymer molecule of length N. Namely, the set of two energy levels of a diatomic molecule becomes a band with N levels.

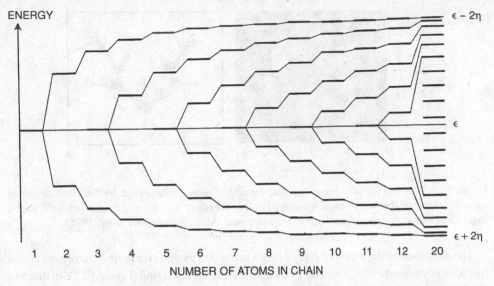

Figure 10.4 Schematic distribution of overlap energies for sets of N atoms each having one s-state.

The number of levels equals the number of atoms for each state. The zero of the energy scale is at the level for one atom. This splits into two levels when the s-amplitude function of this atom overlaps with the amplitude function of another identical atom. One of these is lower (bonding), and one higher (anti-bonding) than the energy of the non-bonded single atom. When another atom is added to the chain ($N = 3$), the two levels split further to create three energy levels (bonded, non-bonded, and anti-bonded). The splitting continues as N increases, with less magnitude at each step so the total range of energies asymptotically approaches a constant value with increasing N, thereby generating a band of energies of width 4η. For even values of N, half the levels are bonding and half anti-bonding.

In the presence of electrons, each state may be occupied by two electrons (one spin up, and one spin down) starting at the lowest unoccupied state, and ranging up to the Fermi energy. The strength of the polymer chain is essentially the same as that of the dimer. It is determined by the magnitude of the hopping integral η which is, in turn, determined by the bond length b.

The case we have considered which allowed only nearest-neighbor hopping leads to one energy band. If hopping between next-nearest neighbors is allowed in addition to the nearest-neighbor hopping, then an additional band of energies appears. The second may, or may not, overlap the first one. If not, there will be an energy gap in which there are no allowed energy levels.

If the chain of N atoms that we have considered is not a closed ring, but is a straight line of finite length, then the atoms near the two ends will have different environments from those in the middle of the chain. This will create energy levels that lie outside of the main band of energy levels. These are known as Tamm, or surface, states. They will be considered when fracture is discussed in Chapter 21.

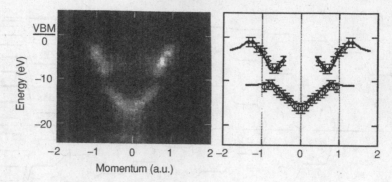

Figure 10.5 Comparison of experimental and theoretical momentum spectra for SiC. Left, scattered intensity showing absence of intensity near a binding energy of 10 eV and gap of about 3 eV. Right, theoretical line with experimental points superimposed. After Vos and McCarthy (1995).

The existence of the bands of energy levels separated by gaps for periodic arrays of atoms has been confirmed experimentally by means of electron scattering (Figure 10.5). In the case of a chain that alternates between two kinds of atoms (such as Si and C in silicon carbide) two bands can be detected with an energy gap between them (Vos and McCarthy, 1995).

References

Atkins, P.W. (1983). *Molecular Quantum Mechanics*, 2nd edn. Oxford: Oxford University Press.

Burdett, J.K. (1995). *Chemical Bonding in Solids*. Oxford: Oxford University Press.

Chen, C.J. (1991). Attractive interatomic force as a tunneling phenomenon, *J. Phys.: Condens. Matter*, **3**, 1227.

Coppens, P. (1997). *X-Ray Charge Densities and Chemical Bonding*. New York: Oxford University Press.

Ferrante, J., Smith, J.R., and Rose, J.H. (1983). Diatomic molecules and metallic adhesion, cohesion, and chemisorption: a single binding-energy relation, *Phys. Rev. Lett.*, **50** (18), 1385.

Feynman, R.P., Leighton, R.B., and Sands, M. (1963). *Feynman Lectures on Physics*, Volume III. Reading, MA: Addison-Wesley.

Keyes, R.W. (1958). An antibonding Morse potential, *Nature*, **182**, 1071.

Kimball, G.E. (1950). *Lecture Notes*. New York: Columbia University.

Morse, P.M. (1929). *Phys. Rev.*, **34**, 57.

Parr, R.G. and Yang, W. (1989). *Density-Functional Theory of Atoms and Molecules*. Oxford: Clarendon Press.

Pauling, L. (1931). *J. Am. Chem. Soc.*, **53**, 1367.

Pauling, L. and Wilson, E.B. (1983). *Introduction to Quantum Mechanics*. New York: McGraw-Hill, 1935. Republished by Dover Publications, New York.

Pettifor, D.G. (1995). *Bonding and Structure of Molecules and Solids*. Oxford: Clarendon Press.

Slater, J.C. (1963). *Quantum Theory of Molecules and Atoms*, Volume 1, *Electronic Structure of Molecules*. New York: McGraw-Hill.

Slater, J.C. (1965). *Quantum Theory of Molecules and Atoms*, Volume 2, *Symmetry and Energy Bands in Crystals*. New York: McGraw-Hill.

Sutton, A.P. (1993). *Electronic Structure of Materials*. Oxford: Clarendon Press.

Vos, M. and McCarthy, I.E. (1995). Observing electron motion in solids, *Rev. Mod. Phys.*, **67**, 713.

11

Intermolecular cohesion

If a neutral atom, or molecule, is placed in an electric field, $\mathbf{E}$, the centers of negative (electrons) and positive (nucleus) charge will separate by some small distance, x. Thus the particle will acquire a dipole moment, $\mu = qx$, where q is the magnitude of the negative charge, if the surrounding medium has a dielectric constant of unity. Such dipoles interact relatively weakly, but always attractively.

For dipoles that are induced, the magnitude of the dipole moment (a vector) depends on the intensity of the applied field, so:

$$\mu_i = \alpha_{ij} E_j \tag{11.1}$$

where α_{ij} is called the polarizability tensor. In one dimension, it reduces to α, and then the charge separation distance is:

$$x = \alpha E / q \tag{11.2}$$

11.1 London forces

From Coulomb's Law, the electric field of a dipole with moment, μ, at a distance, r, from the center of the dipole is:

$$\mu^2 / r^4 \sim \alpha^2 / r^4 \tag{11.3}$$

and its potential energy is:

$$U_{\text{dipole}} = -4\mu^2 / r^{-3} \tag{11.4}$$

Therefore, for two colinear dipoles with parallel moments and with their centers lying a distance r apart (they can be either permanent, or induced, dipole moments; Figure 11.1(a)) the interaction energy is the product of the individual dipole energies:

$$U_{\text{dip-dip}} \sim (U_{\text{dip}}) \times (U_{\text{dip}}) \sim -\alpha^2 / r^6 \tag{11.5}$$

If the dipoles lie along two parallel, but not colinear, lines, the configuration is said to be a quadrupole (Figure 11.1(b)). This configuration has less binding energy than the colinear

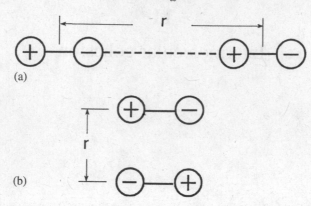

Figure 11.1 Schematic diagrams of dipoles: (a) colinear pairs, (b) pairs forming quadrupoles.

case, and the interaction energy decreases with distance more rapidly (with the eighth power of the distance between the dipoles).

Two molecules (or atoms) in which the average values of x are zero (so the average μ is zero for each one) interact nevertheless by inducing fluctuating dipoles in each other. Each dipole fluctuates because it is a quantized harmonic oscillator, and its lowest energy state has a zero-point energy of $U_0 = h\nu_0/2$, or a minimum frequency ν_0. When the electric fields of the fluctuating dipoles are 180° out of phase, the energy of the pair is reduced, creating an attractive force between the molecules. This is the bound state. When the fluctuating fields are in phase, the energy of the pair is increased and the dipoles repulse one another. This is the anti-bonding state.

An accurate calculation of the bonding and anti-bonding forces requires that the speed with which the electric fields propagate between the dipoles (molecules) and within the dipoles (molecules) be taken into account. If they are close together, this retardation effect introduces only a small correction so it will be ignored initially, but considered in the discussion of Casimir forces.

In each of the dipoles the charge fluctuation frequency, ν_0, is determined by a vibrational force constant, k, and the masses, m, of the dipole:

$$\nu_0 = \frac{1}{2\pi}\sqrt{\frac{k}{m}} \qquad (11.6)$$

From quantum mechanics, the energies of the various oscillation states are:

$$U = (n + 1/2)h\nu_0 \qquad n = 0, 1, 2, \ldots \qquad (11.7)$$

with h Planck's constant. Thus, the ground-state energy ($n = 0$) is $h\nu_0/2$, called the zero-point energy. The existence of this state means that the dipole oscillates even if the temperature is zero, or so small that the thermal energy is too small to excite the oscillations, that is, when $kT \ll h\nu_0$. This is important because otherwise molecular solids would be unstable and vaporize at low temperatures.

In order for a neutral atom, or molecule, to acquire a dipole moment, one or more of its electrons must be excited into a non-spherical state so at least one direction in it becomes unique. For example, one of its electrons might pass from a spherically symmetric s-state into directional p-state. Thus the polarization process is quantized, in principle. However, averaging over several excitations of various magnitudes may make it quasi-continuous in practice.

Consider two dipoles lying a distance, r, apart. For simplicity, let them be identical with charge separation distances, x, and both lie along the x-axis with one of the charges at the origin. Then, the distances between the charges will be r for the two like pairs, $r - x$ for one unlike pair, and $r + x$ for the other unlike pair. The energy of the set is:

$$q^2\{[2/r] - [1/(r+x) + 1/(r-x)]\} \tag{11.8}$$

By neglecting the square terms, this can be reduced to:

$$V = \pm 2q^2 x^2/r^3 \tag{11.9}$$

which is the interaction energy. It is negative for the bonding state, and positive for the anti-bonding state. It acts to perturb the natural frequencies of the dipole oscillators. That is, it adds, or subtracts, from the individual force constants, k, of the dipole oscillators as follows:

$$k', k'' = k \pm 2q^2/r^3 \tag{11.10}$$

or, in terms of frequencies:

$$\nu', \nu'' = (1 \pm z)^{1/2}$$

where $z = 2q^2/kr^3$, and for the ground state ($n = 0$) the energy is:

$$U_0 = (h\nu_0/2)[(1 - z)^{1/2} + (1 + z)^{1/2}] \tag{11.11}$$

If the binomial theorem is applied to the term in square brackets, and terms beyond three are neglected, this becomes:

$$U_0 = h\nu_0(1 - z^2/8) = h\nu_0(1 - q^4/2k^2r^6) \tag{11.12}$$

where the interaction energy is the second term, and it depends strongly on the separation distance r. If this one-dimensional dipole pair is averaged over three dimensions, the factor of $1/2$ in the second term of the equation is replaced by $3/4$. It is instructive to express Equation (11.12) in terms of the polarizability (which is related to the force constant k).

Equating the electrostatic force on the charge, qE, and the restoring force due to the force constant, kx, we find that $x = qE/k$. Combining this with Equation (11.2) yields $k = q^2/\alpha$.

Table 11.1 *Contribution of London force to molecular bonding*

Halogen	Dissociation energy (kcal/mol)	London energy (kcal/mol)	London/ dissociation (%)
F_2	39	4	10
Cl_2	57	20	35
Br_2	45	17	38
I_2	36	13	36

Then the interaction energy of Equation (11.12) becomes:

$$(h\nu_0/2)(\alpha/r^3)^2 \tag{11.13}$$

Numerically, the polarizability is approximately equal to an atomic (molecular) volume, so large atoms, or molecules, tend to be highly polarizable, and therefore to attract one another relatively strongly through induced dipole–dipole forces. These are known as London forces (London, 1930). When r is small (of the order of an atomic, or molecular, radius), Equation (11.13) indicates that the interaction energy is large, but it decreases rapidly as r increases. Therefore, many rather stable molecular dimers, and other associations, that are bound by London forces are observed. Two examples are the dimers of hydrofluoric acid $(HF)_2$ and of acetic acid $(H_3C–COOH)_2$.

To estimate how large the London forces can become consider a pair of molecules with radii 2 Å. Then their polarizabilities will be approximately $(4\pi r^3)/3$, and the distance of closest approach will be $2r$. The frequency, $\nu_0 = [q^2/(4\pi^2 m\alpha)]^{1/2} = 4.3 \times 10^{14}\,s^{-1}$, and $h\nu_0 = 1.88$ eV, so the dipole–dipole interaction energy is one-eighth of this, or 0.24 eV. This is one-tenth, or less, than the intramolecular binding energy. It corresponds to a force constant of about 2.4×10^2 dyn/cm which is about 1000 times smaller than intramolecular force constants. It depends very strongly on the choices of values for α and r, of course. It is very large compared with the gravitational attraction between the two molecules, about 10^{30} times larger.

Pitzer (1955) has shown that London forces play a significant role in the cohesion of many molecules. Table 11.1 lists four halogen molecules together with their dissociation energies, and estimates of the parts of their dissociation energies due to London forces which shorten the bonds. Also given are the fractional effects. The data show that London forces can play a significant role even for covalent bonding, and certainly should not be neglected.

11.2 Polarizability

Polarization consists of changing the shape of a charge distribution by applying an electric field to it, thereby inducing a dipole moment in it. In the most simple case, we start with a spherical distribution of one, or more, valence electrons surrounding a positive ion. Then the applied field changes the shape by shifting the center of the electron cloud relative to the center of the positive ion. As has already been mentioned this cannot be done continuously. It requires the energy state(s) of the valence electron(s) to be changed, typically from an s-state(s) to a p-state(s). This requires transitions from one quantum state to another, that is, from one energy level to another. Thus, we can anticipate that one of the parameters that determines the polarizability is the size of energy difference between the initial and the lowest polarized state. The technique is to consider the change in energy to be small so it can be considered to be a perturbation of the initial state, and perturbation theory can be applied.

Let the initial probability amplitude of the electron distribution be $\psi(x, y, z)$, a solution of the Schrödinger equation. Then the charge distribution is given by $q(\psi\psi^*)$ where q is the charge of the electron, and ψ^* is the complex conjugate of ψ. The Hamiltonian operator, H, acting on ψ gives the energy, U, associated with ψ, times ψ:

$$H\psi = U\psi \tag{11.14}$$

$H = -(h^2/8\pi^2 m)\nabla^2 + V(x, y, z)$, where the first term generates the kinetic energy, while the second gives the potential energy. To obtain the total energy, U, the procedure is to multiply Equation (11.14) by ψ^*, and integrate both sides over the volume $v(x, y, z)$. Then:

$$\int \psi^* H\psi \, dv = U \int \psi^*\psi \, dv = U$$

since U is a constant, and $\int \psi^*\psi \, dv = 1$ because the ψ are normalized.

Now, suppose an electric field, $\mathbf{E}(z)$, is applied. It produces a force on the electron, qE, and hence a perturbation, β, of the potential energy:

$$\beta(z) = qEz \tag{11.15}$$

This perturbation changes the energy operator so $H \rightarrow H + \beta$. It also changes the probability amplitude, ψ, by a small amount, f, so $\psi \rightarrow \psi_0 + f$. And it changes the total energy by a small amount, u, so $U \rightarrow U_0 + u$. With these changes, Equation (11.14) becomes:

$$[(H + \beta) - (U_0 + u)](\psi_0 + f) = 0 \tag{11.16}$$

Dropping second-order terms from this:

$$(H - U_0)f + (\beta - u)\psi_0 = 0 \tag{11.17}$$

The change in the probability amplitude, f, can be written as a sum of correction amplitudes ψ_j that are solutions of Equation (11.14). The coefficients of these correction amplitudes are designated A_j. Therefore, the perturbation of the probability amplitudes becomes:

$$f = \sum_j A_j \psi_j \tag{11.18}$$

and Equation (11.17) is:

$$\sum_j A_j(U_j - U_\text{o})\psi_j + (\beta - u)\psi_\text{o} = 0 \tag{11.19}$$

This is now multiplied by ψ_o^*, and integration over the volume is carried out. Since the ψ are orthogonal, the integral of the first term is zero, and u is a constant, so:

$$u = \int \psi_\text{o}^* \beta \psi_\text{o} \, dv \tag{11.20}$$

Similarly, if Equation (11.19) is multiplied by ψ_j^*, and the volume integration is done, the result is:

$$A_j = \frac{\int \psi^* \beta \psi}{U_\text{o} - u_j} \, dv \tag{11.21}$$

And the probability amplitude for the perturbed system is:

$$\psi(x, y, z) = \psi_\text{o}(x, y, z) - qE \sum_j \frac{z_{jo}}{U_\text{o} - u_j} \psi_j(x, y, z) \tag{11.22}$$

where $z_{jo} = \int \psi_\text{o}^* z \psi \, dv$.

The perturbed probability density is $(\psi \psi^*)$, so the charge distribution is:

$$\rho(x, y, z) = -q(\psi \psi^*) = -q(\psi \psi^*)_\text{o} + q^2 E \sum_j \frac{z_{oj}}{u_j - U_\text{o}} [(\psi^* \psi)_{jo} + (\psi \psi^*)_{jo}] \tag{11.23}$$

neglecting terms of order E^2. Note that the term in square brackets equals $1 + 1 = 2$.

The atomic dipole moment is:

$$\mu = \int \rho z \, dv = \alpha E \tag{11.24}$$

where α is the polarizability. Putting Equation (11.23) into Equation (11.24), the first term in Equation (11.23) (the unperturbed charge density) is symmetric in z, so it integrates to zero. The second term yields:

$$\alpha = 2q^2 \sum_j \frac{|z_{jo}|^2}{h \nu_{jo}} \tag{11.25}$$

Usually just one absorption frequency, ν_{jo}, representing a transition of the electron from the ground state, ψ_o, to the next higher energy state, ψ_1, dominates other absorptions. The ground state might be an s-state, and the excited state a p-state which has a new shape as required. The sum, $\sum_j h \nu_{jo}$, can then be replaced by the dominant energy difference, Δ. Also, the displacement length of the dipole can be expressed in terms of an "oscillator strength", f_{jo}. Thus:

$$|z_{jo}|^2 = \frac{h^2 f_{jo}}{8\pi^2 m \Delta} \tag{11.26}$$

According to the Kuhn–Thomas Sum Rule, $\Sigma_j f_{jo} = N$ which is the number of electrons. Then, for a typical one-electron atom or molecule, the approximate polarizability is:

$$\alpha = \frac{2}{m}\left(\frac{qh}{2\pi\Delta}\right)^2 \tag{11.27}$$

The inverse square dependence of the polarizability on the excitation energy, Δ, indicates that systems (atoms, molecules, solids) with small Δ tend to be easily polarized. That is, they tend to be chemically and mechanically soft. Since they absorb light easily if Δ is small, they also tend to be colored. If Δ is a narrow band as in a dye, the coloration is bright and saturated. Such dyes tend to be quite adherent because of the dipole–dipole attraction. If Δ is a broad band as in some polymers, the color tends to be a muddy brown. Thus, strong adhesives tend to be brownish.

11.3 Casimir forces

For short distances, the exchange of photons between fluctuating dipoles is essentially instantaneous. That is, the time required for a photon to jump from one dipole to the other is short compared with the intradipole vibration periods. For large distances, the time required (called the retardation time) which is the separation distance divided by the velocity of light becomes significant. The first theory of this effect was developed by Casimir (Spruch, 1993), so it is known as the Casimir effect. It is a small effect, particularly for solid mechanics, but it has played an important role in the history of quantum electrodynamics, and it is important in colloid chemistry.

Casimir forces do not become important in comparison with London forces until the transit time between particles becomes greater than the periods of the dipole oscillators, ω_o, that is, for distances, $d \geq c/\omega_o$, or about 1000 molecular diameters (c is the velocity of light).

11.4 Derjaguin forces

Since dipole–dipole interactions are always attractive (unlike ion–ion interactions), collections of dipoles can attract other collections of dipoles. The interactions between the various dipoles are simply additive. In other words, the zero-point fluctuations of the dipoles in one macroscopic block of material can yield a small, but measurable, attraction for the fluctuating dipoles in another macroscopic block. The first reliable measurements of this attraction were made by the Russian physical chemist B.V. Derjaguin (1987), so we shall call it the Derjaguin effect.

Israelachvili (1973), following Derjaguin as well as his own mentor, D. Tabor, studied the interactions of macroscopic solids placed in close proximity. He used his measurements to determine such factors as surface energies.

11.5 Dipole–dipole crystals

Molecular crystals, as well as noble gas crystals, consist of arrays of dipoles. In the most simple cases, each point of a Bravais lattice is occupied by a dipole with its zero-point

Table 11.2 *Some properties of noble gas crystals*

	v_a	α	U_c	U_c'	C_{11}	C_{12}	C_{44}	B	ϵ	σ	d	r_0/σ
Ne	16.6	0.40	0.020	0.037	0.018	0.007	0.006	0.0135	0.0031	2.74	3.16	1.14
Ar	27.6	1.64	0.080	0.125	0.039	0.026	0.020	0.0303	0.0104	3.40	3.76	1.11
Kr	33.5	2.48	0.116	0.169	0.051	0.029	0.027	0.0360	0.0141	3.65	4.00	1.10
Xe	42.8	4.40	0.170	0.240	0.054	0.029	0.028	0.0338	0.0200	3.98	4.34	1.09

v_a atomic volume (Å^3); α polarizability (Å^3); U_c measured cohesive energy (eV/atom); U_c' calculated cohesive energy (eV/atom); C_{ij} elastic constants (Mbar); B bulk modulus (Mbar); ϵ Lennard-Jones binding energy parameter (eV); σ Lennard-Jones collision diameter (Å); d nearest-neighbor distance in crystal (Å). Data from Kittel (1976) and Lide (1999); adiabatic elastic constants from Landheer *et al.* (1976).

energy; and there are collective modes of oscillation of the molecules (or atoms) with the plasmon energy. Cohesion in this case is principally determined by the interactions of each dipole with its nearest neighbors. The most simple case is that of the noble gases Ne, Ar, Kr, and Xe (He is a special case) which form face-centered cubic (f.c.c.) crystals at low temperatures. In these crystals there is little overlap of the atomic electron densities. Their polarizabilities, and other parameters with pertinence to their mechanical properties are listed in Table 11.2. The parameters were measured for these atoms while they were in the gas phase. It will be shown that the gas phase properties determine the solid-state properties because there is so little overlap of the atomic electron densities.

As expected, the polarizabilities in Table 11.2 are proportional to the atomic volumes (Figure 11.2). However, unlike a variety of molecules (Brinck, Murray, and Politzer, 1993) their numerical values are much smaller than the atomic volumes. Note also that the diameters of the noble gas atoms are considerably larger than those of isoelectronic species. Neon, at 3.16 Å, is 36% larger than NaF, at 2.32 Å. Furthermore, a confirmation that the binding is of the dipole–dipole type is that the cohesion increases with the atomic size (Figure 11.3). This is opposite to the behavior of covalent, ionic, and metallic crystals. It results from the binding being proportional to the squares of the polarizabilities which increase with the cubes of the diameters (Equation (11.13)).

Since the polarizabilities are proportional to the atomic volumes (r_0^3), the attractive forces between the noble gas atoms are independent of the atom type at $r = r_0$. Therefore, the repulsive forces that keep the atoms from overlapping determine the differences in the cohesive energies of the crystals. These are sometimes called Pauli repulsive forces because they arise from the Pauli Exclusion Principle. The repulsion increases rapidly with decreasing separation of the atoms below a characteristic distance. This rapid rise can be described using either exponential, or power, functions. The latter are the most popular, although they are less consistent with wave mechanics. Most common of all is the Lennard-Jones repulsion:

$$U_{\text{rep}} = +B(\sigma/r)^n \qquad n = 12 \tag{11.28}$$

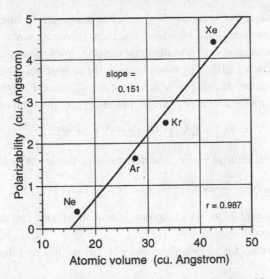

Figure 11.2 Polarizability versus atomic volume for the noble gases.

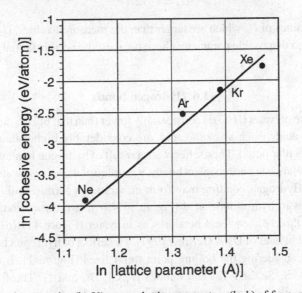

Figure 11.3 Cohesive energies (ln U) versus lattice parameters (ln b) of f.c.c. noble gas crystals.

where B is an energy constant, and σ is a characteristic size. When r becomes less than σ, U_{rep} increases very rapidly. Note that if n tends toward infinity, this becomes a "hard sphere" potential energy.

Combining the attractive (Equation (11.13)) and repulsive (Equation (11.28)) potentials, and redefining the constants, yields a standard form for the Lennard-Jones equation:

$$U(r_{ij}) = 4\epsilon[-(\sigma/r_{ij})^6 + (\sigma/r_{ij})^{12}] \qquad (11.29)$$

where r_{ij} is the distance between any two atoms in a crystal. To obtain the total energy, the energies of all the pairs must be added together. However, since Equation (11.29) is approximate, and the interest here is to illustrate the principles rather than produce a precise result, this task will be simplified by considering only nearest-neighbor pairs. Then, since the structure is face-centered cubic in which there are 12 nearest neighbors for each atom, and there are $N/2$ pairs for a total of N atoms in the crystal, the total energy is:

$$U_{tot} = 2N\epsilon\{12[-(\sigma/r)^6 + (\sigma/r)^{12}]\} \tag{11.30}$$

to find the equilibrium value of $r = r_0$, the derivative of this is set to zero:

$$dU_{tot}/dr = -24N\epsilon[-(6/r)(\sigma/r)^6 + (12/r)(\sigma/r)^{12}] = 0 \tag{11.31}$$

and the solution is $r_0 = (1.12)\sigma$, which agrees quite well with the last column of Table 11.2 giving the "experimental" values.

Substituting the value for r_0 into Equation (11.30) results in values for the cohesive energy:

$$U_c' = -12N\epsilon \tag{11.32}$$

Table 11.2 lists values of U_c' which are larger than the measured values, U_c, but are not bad considering that no disposable parameters have been used, and substantial approximations.

11.6 Hydrogen bonds

The vapor pressure of water (H_2O) is considerably lower than that of H_2S, and the viscosities of liquid organic acids, such as oxalic acid, are considerably higher than expected for a simple monomolecular liquid. These effects are a result of hydrogen bonding. Although this bonding is weak relative to pure covalent bonds, it is ubiquitous because water and its cousins are so common. Hydrogen bonding may form dimers as in formic acid (Figure 11.4(a)), linear polymers as in hydrofluoric acid (Figure 11.4(b)), two-dimensional protein sheets (Figure 11.4(c)), three-dimensional networks as in water (Figure 11.4(d)), as well as the three-dimensional helices of DNA (Jeffrey, 1997). The effect of hydrogen bonds on stability may be seen by considering the boiling point trend lines (Figure 11.5) of the hydrogen chalcogenides (O, S, Se, and Te) and halides (F, Cl, Br, and I). The former trend line suggests that the boiling point of water might be expected to be about $-85\,°C$, compared with the actual value of $+100\,°C$. The latter trend suggests that hydrofluoric acid should boil at about $-120\,°C$, compared with the actual value of $+20\,°C$. The differences of 185 and $140\,°C$, respectively, are striking, and are attributable to the fact that oxygen and fluorine are more electronegative than hydrogen. Therefore the pairs OH and FH form strong dipoles that interact with O and F to form the bound groups OHO and FHF. As a result the molecules in water and hydrofluoric acid become strongly associated, rather than independent. This markedly changes their chemical and physical properties. The only other element that is sufficiently electronegative to form strong hydrogen bonds is nitrogen (Pauling, 1964).

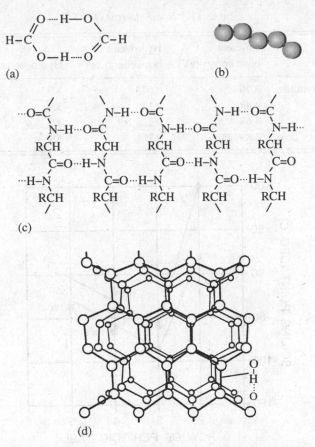

Figure 11.4 Configurations resulting from hydrogen bonding: (a) formic acid molecular dimer, (b) linear hydrofluoric acid polymer, (c) protein sheet, (d) ice, the circles represent the oxygen atoms and the "sticks" are the hydrogen bonds.

The strengths of hydrogen bonds are small compared with covalent bonds, but they are comparable with van der Waals (London) bonds, so they often play an important role in the macroscopic behavior of matter, particularly in connection with the behavior of water near and at its melting point. These bonds give ice its open structure which densifies upon melting. In ice the oxygen atoms are in a tetrahedral array (the same as the carbon atoms in diamond) with hydrogen atoms "hopping" (or exchanging) between the oxygens along the tetrahedral bond paths. Thus the hydrogens are analogs of the electrons that hop back and forth to link the carbons in diamond.

By linking the oxygen and nitrogen atoms of proteins, hydrogen bonds also result in the formation of the helical polymeric fibrils that are so important in biological phenomena.

The relative magnitudes of covalent and hydrogen bonds are indicated by the data in Table 11.3 (Shriver, Atkins, and Langford, 1990). Since one way of viewing hydrogen

Table 11.3 *Bond energies (eV)*

Substance	Covalent bond energy (eV)	Hydrogen bond energy (eV)	Hydrogen/covalent (%)
Hydrogen sulfide	3.76	0.073	1.9
Ammonia	4.47	0.176	3.9
Water	4.68	0.228	4.9
Hydrofluoric acid	5.89	0.301	5.1

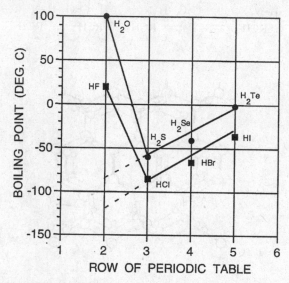

Figure 11.5 Effect of hydrogen bonding on the boiling points of monohydrides and dihydrides. The intersections of the dotted lines with the line for the second row of the Periodic Table indicate the values expected for H_2O and HF in the absence of hydrogen bonding.

bonds is as dipole–dipole interactions, it is not surprising that the magnitudes of the bond energies are similar to other van der Waals energies (Table 11.2).

References

Brinck, T., Murray, J.S., and Politzer, P. (1993). Polarizability and volume, *J. Chem. Phys.*, **98** (5), 4305.
Derjaguin, B.V., Churaev, N.V., and Muller, V.M. (1987). *Surface Forces*, trans. V.I. Kisin and J.A. Kirchener. New York: Consultants Bureau.
Israelachvili, J.N. (1973). *J. Chem. Soc. Faraday Trans. 2*, **69**, 1729.
Jeffrey, G.A. (1997). *An Introduction to Hydrogen Bonding*. New York: Oxford University Press.
Kittel, C. (1976). *Introduction to Solid State Physics*, 5th edn. New York: Wiley.
Landheer, D., Jackson, H.E., McLaren, R.A., and Stoicheff, B.P. (1976). Elastic constants of krypton single crystals determined by Brillouin scattering, *Phys. Rev. B*, **13** (2), 888; see also *Phys. Rev. B*, **11** (4) (1975) 1705; *Phys. Rev. B*, **10** (8) (1974) 3487; *Phys. Rev. B*, **4** (12) (1971) 4518.

Lide, D.R. (1999). *Handbook of Chemistry and Physics*, 80th edn. Boca Raton, FL: CRC Press.

London, F. (1930). *Z. Phys. Chem. B*, **11**, 222; *Z. Phys.*, **60**, 491; see also *Faraday Soc. Trans.*, **33** (1), 8.

Pauling, L. (1964). *College Chemistry*, 3rd edn., p. 455. San Francisco, CA: W.H. Freeman.

Pitzer, K.S. (1955). London force contributions to bond energies, *J. Chem. Phys.*, **23**, 1735.

Shriver, D.F., Atkins, P.W., and Langford, C.H. (1990). *Inorganic Chemistry*, p. 291ff. New York: W.H. Freeman.

Spruch, L. (1993). In *Long-Range Casimir Forces*, ed. F.S. Levin and D.A. Micha, p. 1. New York: Plenum Press.

12

Bulk modulus

Having considered the cohesion within atoms and molecules, and between atoms and molecules, we are in a position to consider the factors that determine the magnitudes of the elastic coefficients connecting stresses and strains. We wish to see how these coefficients are related to the electronic structures, that is, to chemical factors.

The elastic response coefficients are the most fundamental of all of the properties of solids; and the most important sub-set of them is the shear coefficients because they determine the existence of the solid state. If the shear stiffnesses were not sufficiently large, all matter would be liquid like. There would be no aeronautical, civil, or mechanical engineering; and modern micro-electronics, as well as opto-electronics would not be possible. They set limits on how strong materials can be, how slowly geological processes occur, and how natural structures respond to wind and rain. This is why the scientific study of them which began with Galileo (the founder of physical mechanics) continues today. However, the shear moduli are considerably more difficult to interpret in detail than the bulk moduli. They depend on both the shear plane, and the shear direction; and the structures of both of these depend on crystal symmetries and local atomic structures. The bulk modulus, on the other hand, is a scalar quantity relating an isotropic pressure to an average change in volume. It is the average of the three inverse linear compressibilities (change of length induced by pressure). Although the shear moduli have more influence on mechanical behavior, we shall discuss the simpler bulk moduli first.

One way of classifying solids is by means of a *solidity index*. This is simply the ratio of the shear modulus to the bulk modulus multiplied by a coefficient of order unity. This index is zero for a liquid and reaches its maximum value of 1.3 for diamond. If the solidity index S is defined as:

$$S = (3/4)[G/B] \tag{12.1}$$

it runs between 0 and 1. One use for S is to roughly distinguish *ductile* and *brittle* solids. This was originally suggested by Pugh (1954), and developed by Gilman, Cunningham, and Holt (1990). Cottrell (1997) has suggested that a dividing point of $S = 0.23$ is about right, brittleness being associated with $S > 0.23$. This index characterizes a solid much better than the bulk modulus alone which has the same magnitude for many liquids as it does for the corresponding solid. That is, in many cases B is nearly the same on both sides of the melting point, so it characterizes substances overall, but not their solid states explicitly.

Another way to classify solids is by bonding type (metallic, covalent, ionic, and molecular), as mentioned in Chapter 9. The extremes of these are quite distinct: metallic (sodium or aluminum), covalent (diamond or silicon), ionic (potassium chloride), and molecular (argon or polyethylene wax). These categories will be used as major sub-headings here after the general pattern of the elements is described.

12.1 Bulk stiffnesses of the elements (chemical factors)

A major source of information regarding trends in elastic stiffness is the Periodic Table of the Elements. Figure 12.1 is a plot showing the bulk moduli of the elements for most atomic numbers up to 100. In order to improve the "dynamic range" the natural logarithms of the moduli have been plotted. This has allowed variations covering about three orders of magnitude to be shown on one graph. The smallest values (less than ln 0+, in units of megabars) could not be plotted, so they have been arbitrarily placed along the abscissa. The data are from Kittel (1996) and Ledbetter (1983).

It may be seen that there are six peaks in the bulk stiffness corresponding to each of the rows of the standard Periodic Table. The minima fall at the noble gases, and the maxima when the valence shells of rows one and two are half full of electrons; and then when each of the transition shells (3d, 4d, and 5d) is roughly half filled. The peaks are not smooth maxima, but show various deviations from smoothness; especially at 25 (manganese), 31 (gallium), 63 (europium), and 70 (ytterbium).

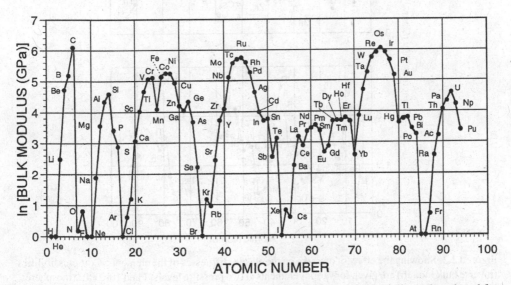

Figure 12.1 Bulk moduli of the elements. Note that the logarithms of the moduli are plotted, and for those elements with moduli less 1 GPa, the logarithm has been arbitrarily set at zero.

In addition to the patterns of Figure 12.1, there are other patterns that relate to the sizes of the ion cores of the atoms, and their net charges (nuclear charge minus the ion core charge). Both of these depend on the number of valence electrons (column of the Periodic Table) as well as the sizes of the ion cores (atomic numbers). The sizes of the ion cores are affected by the amount of screening from the nuclear charge that is experienced by the various electrons depending on their relative positions within the cores.

12.1.1 Effect of pressure on the compressibilities of the elements

At sufficiently high pressures, all of the elements tend to have the same bulk moduli, or the inverses, the compressibilities. Figure 12.2 indicates this by showing the compressibilities of the elements at two pressure levels: 1 bar (1 atmosphere) and 100 kbar. Note that the elements with the maximum compressibilities (the alkali metals) decrease from 2–3 Mbar^{-1} to about 1.2 Mbar^{-1}, and the variation among them decreases. Note also that the width of the band of values for the entire data set decreases by a factor of about three when the pressure is increased from 1 bar to 100 kbar.

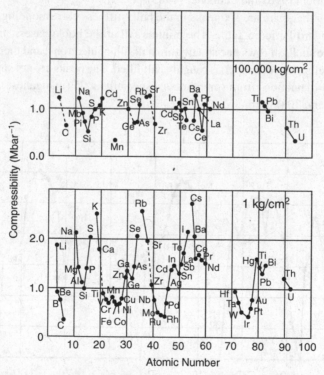

Figure 12.2 Showing the effect of pressure on the elastic stiffnesses of the elements. Compressibilities (inverse bulk moduli) are given for various elements at two pressure levels, 1 and 100 000 atmospheres. Note the considerable reduction in the range of the values. Also, the elements with relatively high compressibilities are affected substantially more than those with relatively low compressibilities.

As the elements become compressed, the valence electron densities in them increase, making them increasingly resistant to further compression. Also, the probability amplitudes of the valence electrons increasingly overlap, so they tend to uniformly become metals in accordance with Herzfeld's criterion (Edwards and Sienko, 1983). At a pressure in the range 10–100 Mbar, all of the elements become metals.

12.2 Metals

Historically, metals have been broadly defined as materials that possess ductility. That is, they are materials that can be forged with hammers, and/or drawn with dies. For a more detailed discussion of the factors that determine ductility see Section VI. In more recent times physicists have defined metals as materials in which electricity is conducted readily by means of electrons (and/or holes). These two definitions constitute a very large category of substances. Consider just the elements. Out of the first 100 elements, nearly 80% are metals in their standard states. Just the binary combinations of these 80 elements form a set of 3160 alloy systems. So it is not surprising that there are several sub-divisions of the general "metal" category. The more prominent sub-divisions will be considered in turn.

There are two principal sources of elastic stiffness in materials with predominantly metallic bonding. One is Coulombic electrostatic forces, and the other is electrodynamic de Broglie forces (also called Schrödinger forces). The bulk modulus is determined almost entirely by the resistance of the outermost, or valence, electrons to compression. These electrons behave like a dense gas, or liquid, with only a very small amount of viscosity. Therefore they do not contribute significantly to the shear stiffness. The latter is determined by the resistance to deformation of the Coulombic interaction between the negative charges of the valence electrons and the positive charges of the atomic nuclei. This is profoundly affected by the Pauli Exclusion Principle.

12.3 Simple metals

Figure 12.3 shows that the bulk moduli of the twenty elements classified as the "sp" elements vary systematically according to their positions in the Periodic Table. The Arabic numbers refer to rows 2 through 6 (row 1, consisting of H and He, is omitted), while the Roman numbers refer to the first four columns, I through IV. Thus, the tallest bar at 2-IV represents carbon (diamond form), while the smallest at 6-I represents cesium metal. The principal point to be learned from the figure is that the bulk moduli of the elements do vary systematically within the Periodic Table. As the number of valence electrons increases, the bulk stiffness increases; and it decreases as the atomic size increases.

The "sp" designation comes from the fact that only the s and the p quantum states contribute to the cohesion of these elements. This distinguishes them from the transition metals in which the occupied d- and f-states play an important role in the cohesion. Up to atomic number 20 (calcium, $1s^2\ 2s^2\ 2p^6\ 3s^2\ 3p^6\ 4s^2$) none of the d-states are occupied in the lowest energy configurations. The heavier elements of Figure 12.3 (Groups III and IV,

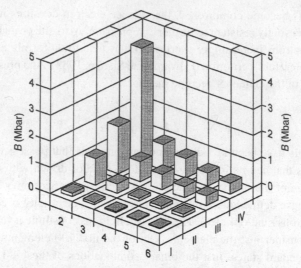

Figure 12.3 Pattern of bulk moduli for columns I–IV and rows 2–6 of the Periodic Table. Position 2-IV corresponds to carbon (diamond), while position 6-I is cesium metal. The number of valence electrons increases from one to four with the column number, and the atomic diameter increases with the row number.

rows 4, 5, 6) do not strictly fit the simple metal category, but they follow the same pattern for the bulk moduli as the lighter elements in the figure.

On the basis of their chemical reactions with various elements, particularly oxygen, the metals in Figure 12.3 have valence states of the "s" and "p" types. The elements in Group I, going from 1 to 5, are Li, Na, K, Rb, and Cs. In Group II they are Be, Mg, Ca, Sr, and Ba. In Group III they are B, Al, Ga, In, and Tl, and in Group IV, C, Si, Ge, Sn, and Pb. The number of valence electrons is equal to the group number. Only the first three of the four groups consist of simple metals, but Group IV is included in order to emphasize the importance of valence electron concentration for the bulk modulus. This will be a recurring theme: stiffness increases with valence electron density. In the figure, it increases systematically with the number of valence electrons per atom in going from Group I to IV, and with the decreasing total electrons per atom (atomic number) in going from row 5 to 1.

The sizes of the atoms change with position in the Periodic Table. This suggests that instead of the number of valence electrons per atom, the volume density should be considered (i.e., the number of valence electrons divided by the atomic volume). Figure 12.4 results. The close correlation over a broad range of values indicates that the principal factor determining the bulk stiffness is simply the valence electron density. Other factors play minor roles.

Four valences characterize the 20 elements of Figure 12.3, one for each of the four elemental groups I, II, III, and IV. The volumes of the atoms have been determined from their crystal structures, and have been used to construct Figure 12.4. A theory of the relationship will be presented shortly.

Further evidence for the importance of valence electron density is provided by the case of Os ($B = 4.2$ Mbar), the highest point except for diamond in Figure 12.1. The atomic

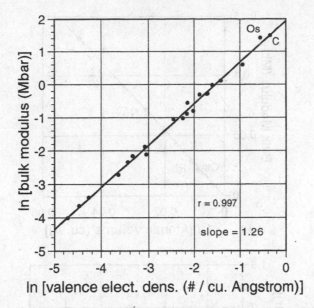

Figure 12.4 Demonstrating that the bulk moduli of the simple elements (sp-bonded elements, groups I–IV) depend principally on their volumetric valence-electron densities. The logarithms of the moduli of Figure 12.3 are plotted against the corresponding logarithms of the valence electron densities. The correlation coefficient of 0.997 indicates that the correlation is excellent. The value from Figure 12.1 for osmium is also plotted to indicate that it is essentially the same as the value for diamond.

volume of Os is about 14.7 Å^3, and it contains six 5d electrons. If they are taken to be its valence electrons, this yields a valence electron density of 0.41 Å^{-3}. For the other high point, diamond ($B = 4.4$ Mbar), the atomic volume is 11.3 Å^3 and there are four valence electrons, yielding a valence electron density of 0.35 Å^{-3} or 15% less than the Os value. Although this may be circumstantial, it is remarkable since the total number of electrons for an Os atom is 76 compared with 6 for carbon. Even if the valence count for Os is in error by ±1, the valence electron density is comparable with that of diamond; so the point for Os falls close to that for C in Figure 12.4.

The mechanical properties of diamond and osmium are quite different. The former is a hard brittle insulator, while the latter is a ductile metal, relatively soft compared with diamond. Thus, in spite of some statements in the literature, the bulk modulus alone does not determine plastic hardness. An important mechanical difference between diamond and osmium is that the shear modulus of diamond is greater than its bulk modulus, while the opposite is the case for osmium.

12.4 Alkali metals

As has been stated previously, since the chemical bonding forces are electrostatic, a correlation of elastic stiffnesses and the Keyes parameter (Keyes, 1962) should be expected.

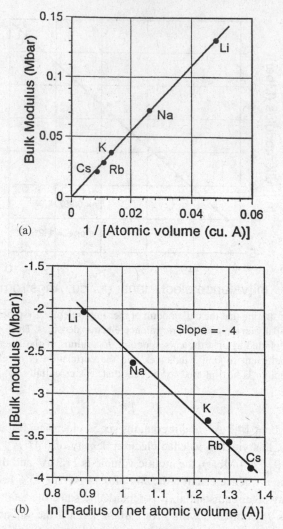

Figure 12.5 Bulk moduli of the alkali metals: (a) abscissa shows uncorrected valence (equal to 1) electron densities; (b) abscissa shows radii derived from the net atomic volumes, equal to total atomic volumes minus ion core volumes. Ln–ln plot showing the power function dependence expected from electrostatics.

Figure 12.5 for the alkali metals considers the behavior. Room temperature data for these metals can be described in terms of the Keyes parameter. However, in recent years low temperature data have become available, and these are used in Figure 12.5. The slope of the regression line for the data of Figure 12.5(a) is about +1.28 rather than the +1.33 expected by Keyes theory. A probable interpretation of the discrepancy is that the volumes of the ion cores of the various atoms should be subtracted from the total atomic volumes to obtain "effective atomic volumes". This procedure removes the discrepancy quite well

Table 12.1 *Ratios of ion core to atomic*
volume for alkali metals

Metal	Ratio (%)
Li	6.3
Na	14
K	20
Rb	23
Cs	27

in the case of the alkali metals as illustrated in Figure 12.5(b). The volumes of the alkali cores are known from X-ray scattering measurements, as are the total volumes (Pauling, 1960), so the radii of the net (or effective) valence volumes can be readily calculated. The figure shows that the bulk moduli are then inversely proportional to the fourth power of the effective radii, as required by the Keyes parameter. The alkali metals are a favorable set because their ion cores are small compared with their total atomic volumes. Table 12.1 shows the volume ratios (ion core/total) ranging from 6% for Li, to 27% for Cs.

This approach is a simplified equivalent of pseudopotential theory. In this case instead of adjusting the potential of an atom to fit experimental measurements, the volume available to the valence electrons is adjusted as indicated by chemical information to fit a physical property.

As a result of their work on the universal binding energy relation (UBER), Rose, Ferrante, and Smith (1981) found, for a large variety of chemical bonding situations, that the most important factor in determining the strengths of bonds is the valence electron density (Banerjea and Smith, 1988). Since B is a measure of bond strength, Figures 12.4 and 12.5 are consistent with the UBER results. The existence of the UBER justifies the inversion of the electron density–bulk modulus correlation to obtain the bulk moduli. The UBER will be described in more detail in Chapter 15.

12.5 Transition metals

There are various other patterns of elastic stiffness throughout the Periodic Table, but there is insufficient space to discuss all of them here. One that is especially important, and will be discussed briefly, is the pattern displayed by the transition elements. These metals have particular importance because they are the main ingredients of many structural as well as magnetic materials, iron being the most important of all.

The transition metals make up the three broad peaks (periods) of Figure 12.1. In terms of atomic numbers they consist of:

transition period one, atomic numbers 21–30
transition period two, atomic numbers 39–48
transition period three, atomic numbers 57–80

In each case the s-state electrons are less important to the bonding than those in d-states, so they are neglected in counting the number of valence electrons.

A standard viewpoint is that as the number of valence d-electrons increases from one to five in each d-band, the bonding increases in strength (Cottrell, 1988). At this point the respective d-band becomes half full. That is, the half of the ten d-orbitals (the bonding orbitals) with energies less than the d-levels in the same atoms at wide separations are all occupied. When the next d-electron is added, it must go into an anti-bonding orbital. This lies in energy above the d-level of the widely separated atoms. Thus the number of *effective* valence d-electrons decreases, and properties like the bulk modulus begin to decrease; and they decrease further as more electrons are added.

This simple rationale has considerable appeal. Unfortunately it is not consistent with the factual data of Figure 12.1 which show that the maximum bulk modulus lies at $z_d = 6$, not at $z_d = 5$. Therefore, this simple model, attributed to Friedel, is unsatisfactory, and will not be pursued further here.

A simple pattern for the stiffnesses of the transition metals in terms of their positions in the Periodic Table will be demonstrated. The basis of this pattern is not known, but it correlates, not all, but a large fraction, of the data. It is based on postulating a simple pattern for the number of valence electrons, calculating valence electron densities from these numbers, and then plotting the measured values of the bulk moduli versus the postulated electron densities. Surprisingly enough, this yields a universal linear regression line. It is assumed that the s-state electrons do not contribute significantly to the bulk stiffnesses.

The pattern of d-electron valence numbers chosen for the third long period, for example, is the following symmetric series (with the valence numbers in parentheses):

La(1), Hf(2), Ta(3), W(4), Re(5), Os(5), Ir(4), Pt(3), Au(2), Hg(1)

and similarly for the second and first long periods. The only justification for using this pattern is that it yields the best correlations. These numbers, designated d'-valences, have been divided by the corresponding atomic volumes to give the d'-densities. The only purpose of the primes is to remind the reader that the numbers are somewhat arbitrary.

Figures 12.6, 12.7 and 12.8 are log–log plots of the d'-electron densities, and their corresponding bulk moduli. Except for the magnetic elements, Cr, Mn, Fe, and Co, in Figure 12.6, the plots are linear with high correlation coefficients. All three of the plotted lines lie close together as shown in Figure 12.9 where they are replotted together with the trend line from Figure 12.4 for the sp-bonded metals. The latter are about an order of magnitude smaller than those for the transition metals for any particular valence electron density.

These plots seem to justify the neglect of the s-electrons in calculating the valence electron densities. However, there is no rationalization for them.

For a more comprehensive (but inconclusive) discussion of bonding in the transition metals, the reader is referred to the books by Cottrell (1988), Pettifor (1995), and Sutton (1994).

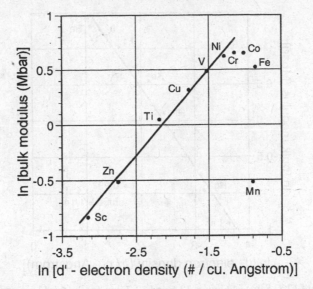

Figure 12.6 For the elements of the first long period, the logarithm of the bulk moduli versus the logarithm of the d′ electron density. The effective number of d-electrons acting as valence (bonding) electrons is d′. The magnetic elements (Cr, Co, Fe, Mn) are exceptions. d′ follows the sequence 1, 2, 3, 4, 5, 5, 4, 3, 2, 1.

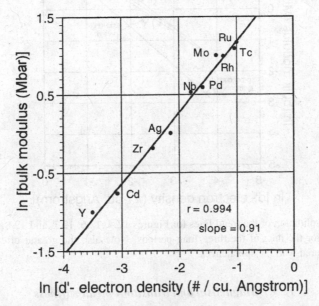

Figure 12.7 Same as Figure 12.6, but for the second long period. Note the high value of the correlation coefficient, $r = 0.994$.

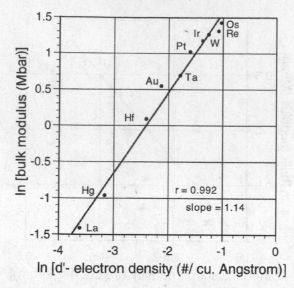

Figure 12.8 Same as Figures 12.6 and 12.7, but for the third long period.

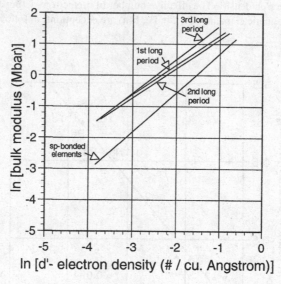

Figure 12.9 Consolidation of the trend lines for Figures 12.4, 12.6, 12.7, and 12.8. Note the nearly universal trends for the data of the three long periods. Note also the greater effectiveness of the d-electrons compared with the sp-valences.

12.5.1 Magnetism and transition metal stiffness

The d-orbitals play a critical role in the transition metals but they do not usually act alone. They become hybridized with the p- and s-orbitals. The evidence of their importance is that as one proceeds across one of the long rows, the atomic volume changes relatively

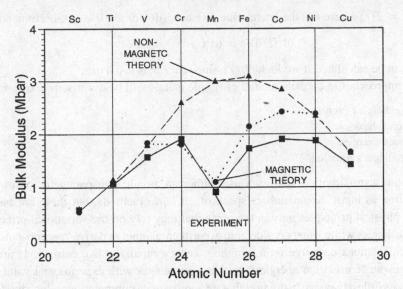

Figure 12.10 Effect of magnetism on the stiffnesses of the elements of the first long period.

little but the elastic stiffness changes markedly, reaching a maximum near the middle of each row. Manganese is an exception in which the bonding is strongly affected by its antiferromagnetism. Other elements that are strongly influenced by their magnetism, but to a lesser extent than Mn, are Cr, Fe, and Co. Figure 12.10 shows the difference between the magnetic and non-magnetic cases as calculated by Moruzzi and Marcus (1993). Although the calculations are approximate, they indicate the importance of magnetism for these elements.

12.6 Theory of the bulk modulus (simple metals)

The bulk modulus is designed to measure the resistance of a material to volume changes under an applied pressure, p. If $(-dV/V)$ is the fractional volume change (compression), the bulk modulus, B, is defined as:

$$B = -dp/(dV/V) \tag{12.2}$$

where p is a positive pressure, and (dV/V) is the fractional reduction in volume. The associated change in the internal energy U (at low temperatures) is:

$$dU = p\,dV \tag{12.3}$$

so:

$$d^2U/dV^2 = dp/dV \tag{12.4}$$

or, if $U = f(r)$ where r is the interatomic distance, $(dU/dr) = 0$ at equilibrium, so:

$$d^2U/dV^2 = (d^2U/dr^2)(dr/dV)^2 \qquad (12.5)$$

and B can be calculated if we know $U(r)$, since dV/dr is just $4\pi r^2$.

Four approaches to the bulk moduli of simple metals will be discussed. They are:

(1) Heisenberg's Principle,
(2) chemical hardness,
(3) plasmon energies,
(4) Schrödinger's equation.

Each is an approximation, and each uses either an atomic size parameter or an energy distribution as input. Some authors speak of *ab initio* methods, but these are based on general physical principles only in the sense that they rely on the variational principle of wave mechanics while using considerable experimental input to derive "pseudo-potentials" so the calculations converge on a minimum energy rapidly. Also, corrections are often applied using "correlation energies" to obtain agreement with experimental values. This author sees little advantage in doing elaborate approximate computations which tend to lose sight of measurement patterns. It is preferable to stay close to the experimental measured quantities.

12.6.1 Heisenberg's Principle applied to simple metals

A simple approach to the bulk modulus can be made through Heisenberg's Theorem using results from Chapter 8. When the macroscopic volume changes, equal volume changes must be experienced by all of the atoms (assuming no phase changes). Thus we only need to consider a single spherical atom of radius R, taking its behavior to be representative of the whole array of atoms. Using Equation (8.8) as shown in Chapter 8, the kinetic energy T of the valence electron in such an atom is:

$$T = 9h^2/32\pi^2 m R^2 \qquad (12.6)$$

Notice that the kinetic energy varies inversely with R^2. Also note that it is positive and therefore repulsive, tending to cause the electron cloud to expand. The attractive potential of the positive ion which decreases with distance less rapidly keeps the cloud from expanding indefinitely.

The electrostatic potential energy has two parts: one is the interaction of the electron with the positive ion; and the other is the self-interaction associated with repulsions between charged elements at different locations within the uniform electron distribution. The first component is $-3q^2/2R$, while the latter is $+3q^2/5R$. Together they provide the total electrostatic potential energy:

$$V = -9q^2/10R \qquad (12.7)$$

The sum of the kinetic energy (Equation (12.6)) and the potential energy (Equation (12.7)) yields $U(R)$, so with Equations (12.2) and (12.5), an expression for B can be found:

$$B = 3q^2/40\pi(R_0)^4 \quad (\text{dyn/cm}^2) \quad (12.8)$$

where $q = 4.8 \times 10^{-10}$ esu, and R_0 is the equilibrium atomic radius (cm).

It has been shown elsewhere that this simple theory agrees with measurements quite well for the monovalent, divalent, and trivalent metals (Gilman, 1971). Thus the stiffness and compressive stability of metals can be interpreted to be a result of the Heisenberg Principle.

Notice that this method uses experimental values of the atomic spacings instead of calculating them. Methods based on solutions of Schrödinger's equation calculate the spacings from pseudopotentials that represent the ion cores of the atoms. However, since accurate experimental values are readily available, it seems sensible to use them, thereby greatly simplifying calculations.

Some authors assert that the bulk modulus should vary with the inverse fifth power (rather than the fourth power) of the atomic spacing, for example Harrison (1980). However, this is based on the stiffness of a free-electron gas which gives an incorrect result because a modulus is not a pressure. It depends on the second derivative of an energy with respect to displacement (not on the first derivative). Thus it depends on both Coulombic attraction of the electrons to the ion cores of atoms, and the Heisenberg–Pauli repulsion.

12.6.2 Chemical hardness

The second method for calculating B for the alkali metals is based on chemical hardness μ (Atkins, 1991). This parameter measures the stabilities of chemical species (Pearson, 1997). It is defined as the change in chemical potential with a change in the number of electrons in an atom or a molecule. In theoretical discussions it is treated as a continuous variable, although in practice it is a discrete quantity (Parr and Yang, 1989). In particular, for an atom, it is the difference between the ionization energy and the electron affinity energy (often with an arbitrary divisor of two). For a molecule, it is half the difference between the energy of the LUMO and the energy of the HOMO (LUMO, lowest unoccupied molecular orbital; HOMO, highest occupied molecular orbital). For a solid, the chemical hardness is one-half of the difference between the conduction-band minimum energy and the valence-band maximum energy (i.e., the minimum band gap). From the definitions, if the bonding electrons in an atom, molecule, or crystal are excited with an energy equal to twice the chemical hardness, the atom, molecule, or crystal becomes unstable. The instability is with respect to internal changes such as a change of shape, crystal structure, or electronic state (metallization) (Burdett, 1995).

For the alkali metals, since the electron affinity energy is very small, the chemical hardness equals half the ionization energy to a very good approximation. If Z^* is the effective nuclear charge (the screened charge seen by a valence electron), while Z is the atomic number, then the ionization energy is given approximately by:

$$[(Z^* + 1) - Z](q^2/R) \quad (12.9)$$

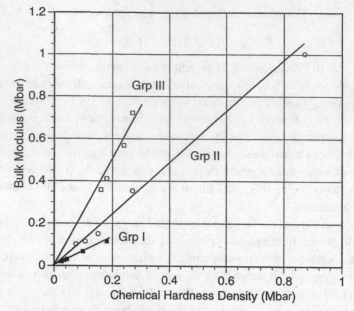

Figure 12.11 Relationship between chemical hardness and mechanical stiffness. Both depend on valence electron density. In the figure, Grp I refers to the alkali metals, Grp II to the alkaline earth metals, and Grp III to the aluminum column metals.

Parr and Yang (1989) have shown that the chemical hardness, μ, is proportional to $(1/R)$. Then, using Equation (12.8):

$$B = A(\mu/V) \tag{12.10}$$

where A is a constant and V is the volume (Gilman, 1997). So B is expected to be proportional to the chemical hardness density which for the alkali metals is simply the first ionization energy. Figure 12.11 confirms the validity of Equation (12.10) for the three groups of simple metals: the lithium group (I), the beryllium group (II), and the aluminum group (III).

12.6.3 Plasmons

When beams of moderately energetic electrons (5–40 eV) pass through thin foils of solids it is found that an absorption resonance occurs near the primary beam energy (sometimes a series of resonances uniformly spaced in energy). This absorption peak has been identified with the creation of *plasmons* which represent collective oscillations of the valence electrons. These oscillations are analogous with the polarization oscillations that occur in atoms and molecules as a result of their polarizabilities. In a metal where the valence electrons are nearly free, plasmons propagate through the electron gas at the speed of longitudinal sound waves. Since the viscosity of the electron gas is very small it does not support shear strains, so the speed of the longitudinal waves is related to the bulk modulus of the electron gas. This is one component of the overall bulk modulus (Pines, 1963). Thus, the overall value of B is related to the plasmon frequency, and through Planck's equation to the plasmon energy.

A simple relationship yields the plasmon frequency. Although the plasma oscillations are collective modes, they can be represented as "quasi-particles". These particles are given a charge equal to that of one electron. Then the equation of motion of such a particle can be written in the standard form for a harmonic oscillator. It is written as the sum of three forces: the inertial force, a damping force, and a driving force. Its longitudinal displacement is $u(x)$, its velocity is du/dt, and its acceleration is d^2u/dt^2. It has an effective mass m (which often is just the free electron mass), and a charge q. It responds to an electric field, $\mathbf{E}$, and a damping constant, Γ. Its equation of motion is:

$$m(d^2u/dt^2) + m\Gamma(du/dt) - q\mathbf{E} = 0 \qquad (12.11)$$

For a nearly free-electron gas, Γ is small so the damping term can be neglected. Then, for an oscillatory disturbance such as an incident electron, or photon, with frequency ω, the medium responds by becoming polarized as the nearly free electrons are slightly displaced from their equilibrium positions. For one electron the dipole moment becomes qu, so for an electron concentration of n (number/volume), the polarization $\mathbf{P} = nqu$. From electrostatics, $4\pi\mathbf{P} = -\mathbf{E}$, and the last term in Equation (12.11) becomes $+4\pi nq^2u$. Thus it is the equation of a simple harmonic oscillator, and the solution is $u = u_0e^{i\omega t}$. Substituting this into Equation (12.11) yields $-m\omega^2 + 4\pi nq^2 = 0$. Here ω is called the plasma frequency, ω_p, because it is analogous with the resonant frequencies observed in gaseous plasmas. It can be expressed (cgs units) as:

$$\omega_\mathrm{p}^2 = 4\pi nq^2/m \qquad (12.12)$$

The quasi-particle, or plasmon, has the energy $h\nu_\mathrm{p}/2\pi = \hbar\omega_\mathrm{p}$.

Since photons with energies greater than $\hbar\omega_\mathrm{p}$ excite plasmons in metals, whereas photons with less than this amount of energy do not, the plasma frequency can be measured by determining the optical "transparency edge" which occurs at far ultraviolet wavelengths for the alkali metals (Figure 12.12). A connection is thereby established between the optical and the mechanical properties of solids.

Plasmon energies have been measured for a wide variety of materials including metals, insulators, and semiconductors. Both the optical absorption and the electron energy-loss methods have been used. The latter is a standard electron microscopy procedure. The simple form of Equation (12.12) is obeyed remarkably well using integral numbers of valence electrons for n, the usual electron charge for q, and the mass of a "free" electron for m.

Equation (12.12) implies the existence of a spring-like constant (elastic stiffness) associated with plasmons. Its magnitude is $4\pi nq^2$. It is not necessarily isotropic so it may be associated with linear compressibility, or with bulk compressibility. Since the electrons in a solid are tightly coupled to the ions, a disturbance in the electron structure creates a

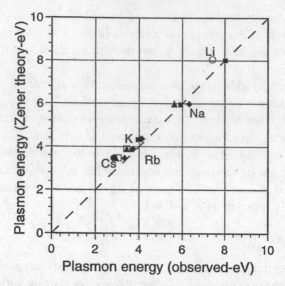

Figure 12.12 Demonstration of the excellent agreement between Zener's theory of plasmons and observations. These values correspond to the transparency edges for the alkali metals. For higher photon energies (shorter wavelengths) the metals become transparent. ■ Kittel, 1996, p. 274; ● Wood, 1933; ▲ Whang, Arakawa, and Callcott, 1972; ◆ Aryasetiawan and Karlsson, 1994; □ Smith, 1970; ○ Gibbons *et al.*, 1976.

disturbance in the ion structure. The latter are sound waves, and indeed there is a close connection between plasma waves and longitudinal acoustic waves.

For the simple metals, as well as the alkali halides and semiconductors, there is an excellent correlation between plasmon frequencies and bulk moduli (Figure 12.13). This can be accounted for quantitatively through Equations (12.8) and (12.10) (Gilman, 1999).

12.6.4 Schrödinger's equation

Since the bulk modulus has spherical symmetry by definition, we use the radial form of the Schrödinger equation:

$$\frac{d^2\psi}{dr^2} + \left(\frac{2}{r}\right)\frac{d\psi}{dr} + \left[\frac{2m}{\hbar^2}(U - V) - \frac{l(l + 1)}{r^2}\right]\psi = 0 \qquad (12.13)$$

In the case of the alkali metals, this simplifies considerably. The single valence electron for each of these is in an s-type orbital which is spherical so there is no angular momentum ($l = 0$). Also, as has already been pointed out, the positive ion cores are small compared with the sizes of the s-orbitals, so the charge density in the equivalent atom (the Wigner–Seitz cell) is nearly uniform. This means that $\psi\psi^\star \approx \psi^2 \approx$ constant, so both of the derivatives with respect to r are zero, and all that remains is $U = V$. This models a single alkali atom, but experimental measurements of the properties of alkali metal crystals indicate that assuming

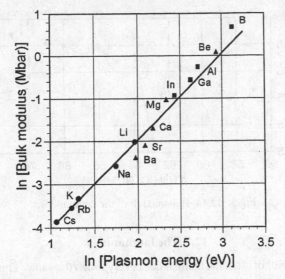

Figure 12.13 Showing that bulk moduli are closely related to plasmon energies for the simple metals. (● Group I, ▲Group II, ■ Group III, line slope of 2). Hence the mechanical and optical properties of metals are linked.

that the solid is the sum of discrete atoms is a good approximation. For example, the discussion above of the connection between chemical hardness and bulk moduli indicates this.

Applying Equations (12.5) and (12.7) to the result that $U = V$ where $V = -0.90q^2/r$, and the Wigner–Seitz cell radius r_s is 1.14 times the atomic radius for r, yields the following expression for the bulk modulus:

$$B = 0.089\left(q^2/\pi r_s^4\right) \tag{12.14}$$

If this is compared with Equation (12.8) where the coefficient is $3/40 = 0.075$, it may be seen that it is essentially the same. Thus the use of the Heisenberg Theorem is further validated. Remember that we are using r_s as an input rather than trying to calculate it.

Notice that the effective mass of the electron is missing from Equations (12.8) and (12.13). This implies that it is constant for the simple metals, and is consistent with the fact that the observed plasmon energies for the simple metals agree with calculations if it is assumed that the effective mass is constant. However, patterns of behavior for the conductivities (electrical and thermal) of these metals suggest that their effective masses vary from one metal to another. This leads to Harrison's (1980) position that the bulk moduli should be proportional to the inverse fifth power of r_s. However, the preponderance of experimental evidence and theory supports the inverse fourth power case. Thus, both the electron charge and mass are essentially constant through the simple metal series. In physical theories, as contrasted with chemical theories, the valence numbers are often taken to be fractional. But again, the systematics of the bulk moduli suggest that the valence numbers tend to be integral.

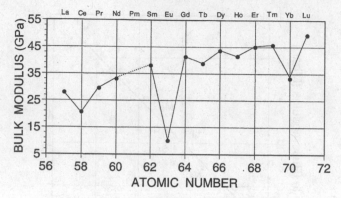

Figure 12.14 Bulk moduli of the lanthanides.

12.7 The lanthanides

The chemical behavior of all of the lanthanides is very much the same. This is to be expected since they all have two 6s valence electrons lying outside shells containing various numbers of 4f-electrons (four of them, La, Ce, Gd, and Lu, also have 5d-electrons) held tightly by their Xe ion cores. The similarity in chemical behavior is reflected in their elastic properties. Their bulk moduli (their shear moduli follow a similar pattern) are shown in Figure 12.14. These data come from Gschneidner and Eyring (1978). There is a trend of increasing bulk stiffness with increasing atomic number with anomalies at Ce, Eu, and Yb which are not understood.

12.8 Hard metals (metalloid–metal interstitial compounds)

These substances consist of relatively large metal atoms usually stacked to form one of two kinds of holes (interstices). The metalloid atoms fit into the holes, and the compounds are sometimes called "interstitial" compounds. Both kinds of hole have coordination numbers of six.

One kind is the octahedral type (close-packing of the larger atoms); the other is the trigonal prismatic type (simple hexagonal stacking of the larger atoms). The prototype of the first type is titanium carbide (TiC) which has the sodium chloride structure so the titanium atoms occupy one of the interpenetrating f.c.c. arrays, while the carbon atoms occupy the other f.c.c. array (i.e., the octahedral holes).

For the second type there are two cases: tungsten carbide (WC) in which the metal atoms are in a simple hexagonal array with half of the interstices filled; and the second case is the diborides, such as titanium diboride (TiB_2), in which all of the interstices are filled with the smaller atoms. In these compounds, the bulk moduli are higher than in the corresponding pure metals. The increases are related to the increased density of valence electrons. Let us consider TiC, for example. The lattice parameter is 4.32 Å, and there are four molecules per unit cell, so the molecular volume is 20.2 $Å^3$. Titanium has four valence electrons, and carbon has four valence electrons, so the valence electron density is 0.40 Å^{-3}. According to Figure 12.4, this corresponds to a bulk modulus of about 2.23 Mbar. The measured value is 2.42 Mbar (Gilman and Roberts, 1961). Thus the difference is about 8%, and the prediction from the valence electron density is quite good.

12.9 Intermetallic compounds

More than 80 of the known chemical elements are metals and many of them interact as pairs, triads, and so on, to form compounds. Therefore, a very large number of intermetallic compounds has been made, and many more must exist. Clear, systematic trends among the elastic stiffnesses are not easily found. The most general is the correlation between valence electron density and stiffness. For discussion of others see Nakamura (1995).

12.10 Covalent crystals

The bulk modulus of diamond is unusually high, giving diamond a special place in the hierarchy of mechanical properties. The reason for this is the small ion core that underlies its valence electrons. Figure 12.15 illustrates this by comparing the valence electron densities of the Group IV crystals. The valence electron densities (number/Å^3) are plotted versus the atomic numbers (total electrons/atom). It may be seen that at silicon there is a distinct break in the trend line. The valence electron density for diamond is more than three times what it would be if the trend line for Si, Ge, Sn, and Pb were followed. In crystals with covalent bonding, charge tends to become localized between neighboring pairs of atoms (Figure 9.3). Unlike the metallic and ionic bonds, these electron charge concentrations lie

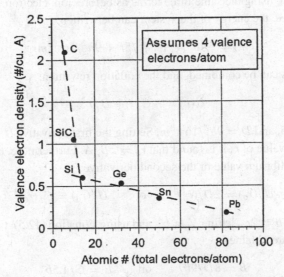

Figure 12.15 Indicating the marked difference between the valence electron density of carbon and the other elements of Group IV. This partially accounts for the uniqueness of diamond.

in definite directions. From a given atom, they may lie in two, three, four, six, or eight directions. Therefore, they may link up atoms into chains (e.g., poly-sulfur-nitrogen), or into two-dimensional meshes (e.g., graphite), or three-dimensional cells (e.g., silicon).

Whereas metallic bonding tends to minimize volume changes, covalent bonding tends to minimize shape changes. In other words, changes in the angles between covalent bonds are strongly resisted. This leads to high hardness, and to brittleness. This will be discussed further in the next chapter.

Covalent bonds arise because the atomic probability–amplitude functions (like other wave-like functions) can interfere with one another either constructively (bonding), or destructively (anti-bonding). In the time-independent case, when the interference is constructive, the average charge density is increased between the atoms, while the kinetic energies of the electrons are decreased because they become spread over both atomic volumes, and therefore less localized (McWeeny, 1990). Equivalently, bonding may be viewed as a dynamic process in which the electrons tunnel (hop) from one atom to the other at a rate that increases as the atoms become closer to one another, until they begin to overlap. When they tunnel in phase, the average charge concentration between the atoms is increased, and there is bonding. If they tunnel out of phase, the opposite occurs, and there is anti-bonding (Chen, 1991).

An estimate of the bulk modulus can be made using an approach that is quite similar to the one used for the simple metals, except that the geometric configuration is different. It belongs to a class of models known as "bond charge models" (Parr and Yang, 1989). Let b be the bond length, and let the two bonding electrons uniformly fill a spherical space between the positive ions. Thus there is a sphere of radius $b/2$ with two electrons in it, and it represents an electron pair bond. At diagonally opposite positions on its perimeter lie the positive ions. Then, using the same three terms as before, ion–electron, electron–electron, and kinetic energies, the energy of the system can be written:

$$U(r) = -4q^2/r + 3q^2/5r + 9h^2/16\pi^2 m r^2 \tag{12.15}$$

The first two terms can be combined, and the equation rewritten:

$$U(r) = -C/r + D/r^2 \tag{12.16}$$

where $C = 17q^2/5$, and $D = 9h^2/16\pi^2 m$. Setting the first derivative $U'(r_0) = 0$, where r_0 is the equilibrium value of r, it is found that $r_0 = 2B/A$. Furthermore, there are two ways of writing the equilibrium value of the second derivative:

$$U''(r_0) = 2D/(r_0)^4 \quad \text{or} \quad U''(r_0) = C/(r_0)^3 \tag{12.17}$$

Remembering that $b = 2r_0$, letting $V = b^3$, and using Equation (12.5), two expressions for the bulk modulus are obtained:

$$B = 8D/9b^5 \quad \text{or} \quad B = C/4.5b^4 \tag{12.18}$$

Note the different exponents of b in the two cases. The first expression assumes that the

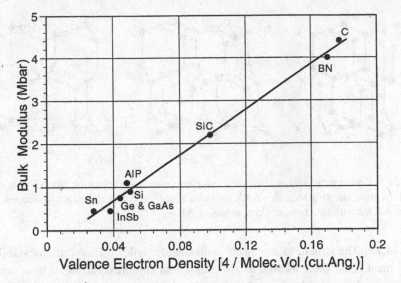

Figure 12.16 Bulk moduli of the homopolar semiconductors and isoelectronic III–V compounds. Their correspondence with simple theory is demonstrated.

electron's effective mass is a constant, while the second assumes that the effective charge is constant. The second contains the Keyes parameter.

Figure 12.16 shows that the form of Equation (12.18) is followed to a good approximation by all of the isoelectronic covalent crystals of average valence 4. The slope of the trend line is close to the predicted unity. The data also agree approximately with the absolute value of C (Equation (12.16)). For diamond, for example, $b = 1.54$ Å, and Equation (12.18) gives $B = 310$ GPa compared with the measured value of 400 GPa. This equation is not expected to cope with the details of the stiffnesses of covalent crystals, but it does give some sense of the physical origins of these stiffnesses. Note that Harrison (1980) favors the "D" version of Equation (12.18), but obtains less good agreement with experiment.

If the atoms of a pair are different, then the bonding charge is not expected to be centered at the mid-point of the pair. It shifts toward the more electronegative atom of the pair. Then the bond is said to be partially ionic. If the shift is nearly complete, as between a strongly electropositive atom and a strongly electronegative one, the bond becomes nearly fully ionic. In this limit, the valence charge is no longer located between the atoms, but is centered on one of the atoms, while a deficit of charge is centered on the other one.

The experimental splitting of the bond charge distribution (Figure 9.3) is due to electron correlation. It plays an important role in shear stiffness and will be discussed in the next chapter.

In the solid state, since eight valence electrons complete a shell, and each covalent bond consists of a pair of electrons, there are four bonds per atom, and each atom is coordinated with four other atoms. This limits the crystal structure to one of three: diamond, zinc blende, or wurtzite. In these three structures the bonds are localized between the atoms

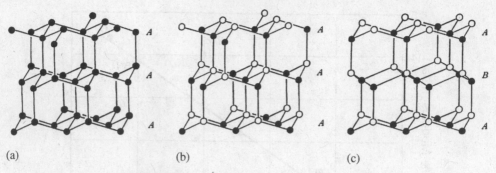

(a) (b) (c)

Figure 12.17 Crystal structures of tetrahedral covalent crystals, coordination number is 4: (a) cubic diamond structure, stacking sequence AAA; (b) cubic zinc blende structure, stacking sequence AAA; (c) hexagonal wurtzite structure, stacking sequence ABA.

(Figure 12.17). The structures are "open" with sizable "holes". Therefore, for stability, the "bonds" must resist shear as well as elongation and compression. In all three structures each atom is surrounded by four others lying at the corners of a tetrahedron. This is the most efficient of all possible open structures since at least four, but not more than four, struts are needed to fix the position of a point in three-dimensional space. The tetrahedral structure also requires a minimum of amplitude functions for the formation of a stationary three-dimensional quantum state. Just two types of function (s and p), and four functions altogether (s, p_x, p_y, p_z) are needed as we showed in Chapter 7, following the brilliant analysis of Linus Pauling.

The three structures of Figure 12.17 are closely related to one another. For the first two, 12.17(a) and 12.17(b), the struts and nodes are arranged in exactly the same pattern, but only one kind of atom occupies the nodes in the diamond structure, whereas neighboring nodes are differently occupied in the zinc blende structure. In Figure 12.17(c) the struts are arranged differently, although there are still four tetrahedral struts for each atom. The difference is as follows. Imagine a shortened "hat tree" with three "legs" arranged at tetrahedral angles, a vertical "trunk", and three "arms" at tetrahedral angles. Then, for the diamond (and zinc blende) arrangement, the arms lie vertically up from the legs; while for the wurzite arrangement, the arms are rotated 60° so they lie vertically up from the bisectors of the angles formed between the legs.

The spring (force) constants between carbon atoms have been measured for a large variety of carbonaceous molecules. They are related to the masses of the carbon atoms, and the vibrational frequencies. For the stretching vibrational mode between two carbons, the force constant is $k_s = 440$ N/m. Since this is related to a volume change, it is of interest to relate it to the bulk modulus B_k which is given by:

$$B_k = k_s/3a$$

where a is the lattice parameter, $a = (16/3)^{1/2}b$, and b is the bond length, $b = 1.54$ Å, so $a = 3.56$ Å, and $B_k = 4.12$ Mbar, compared with the value measured ultrasonically, $B_m = 4.42$; the ratio of these two values is 1.07.

12.11 Ionic crystals

In contrast with localized covalent bonding, ionic bonding is delocalized, and the structures are densely packed. That is, the bonding is distributed over distances that are large compared with the distance between any individual pair of atoms. The most common crystal structure is that of sodium chloride in which each negatively charged atom is surrounded by six positively charged atoms (Figure 12.18(a)). A less common structure is that of cesium chloride in which eight ions of one sign surround each one of the opposite sign (Figure 12.17(b)).

The atoms on the left-hand side of the Periodic Table (the alkali metals) are relatively easily ionized, becoming positive ions. They are said to be electropositive. Those on the right-hand side (the halogen gases) have an affinity for charge, forming negative ions. They are said to be electronegative. The atoms of Group II (the alkaline earths) are less electropositive than the alkalis; and those of Group VI (the chalcogenides) are less electronegative than the halogens. Thus the atoms become less electropositive from one group (column) to another from left to right across the Periodic Table. They also become more electropositive as the rows of the table progress from I to VI.

In accordance with Coulomb's Law, the ions with opposite charges attract one another, while those with like charges repel each other. The closeness of approach of the attractive

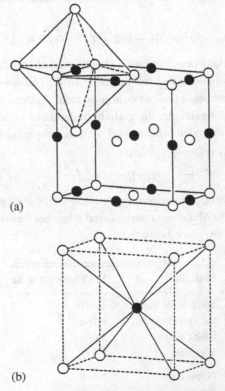

(a)

(b)

Figure 12.18 Structures of highly ionic crystals: (a) sodium chloride structure, coordination number 6; (b) cesium chloride structure, coordination number 8.

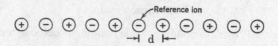

Figure 12.19 String of alternating + and − ions with spacing d.

pairs is limited because the underlying ion cores repulse one another strongly as a result of the Pauli Exclusion Principle. This repulsion arises because of the overlap that occurs as electrons in the filled shells of one ion enter the space occupied by another. A strong resistance to the overlapping occurs because the Pauli Principle forbids more than one electron of each spin to occupy the same quantum state (the same amplitude function). The Pauli overlap repulsion results from the additional constraint that the Pauli Principle imposes on the de Broglie effect.

The long-range nature of the bonding in ionic crystals can be demonstrated by considering a one-dimensional periodic string of positive and negative ions spaced a distance d apart (Figure 12.19). The coordinate origin is placed at one of the negative ions, and then the electrostatic energy of the infinite chain is calculated (Kittel, 1953).

The first interaction to the right of the origin is $-q^2/d$, the next is $+q^2/2d$, then $-q^2/3d$, and so on. Since the same terms appear to the left of the origin, all of the terms must be doubled. Also note that all of the other interactions cancel one another out. Thus, the electrostatic energy is:

$$U_e = -(q^2/d)2[1 - 1/2 + 1/3 - 1/4 + 1/5 \cdots]$$
(12.19)

The term in square brackets converges very slowly toward ln 2 (after ten terms the error is still about 12%) which means that full cohesion of the chain exists only if the chain is very long. Locally, the cohesion is weak or non-existent. In three dimensions the situation is similar. That is why, for example, the grain boundaries in ionic solids tend to be weak (Gilman, 1966). For an extended ionic crystal with nearest-neighbor interionic distances, r, the electrostatic energy is:

$$U_e(r) = -\alpha q^2/r$$
(12.20)

where α represents the three-dimensional summation series analogous to the one in Equation (12.19). It is called the Madelung constant, and it has been evaluated for various crystal structures. A few values are given below:

Structure	Madelung constant
Sodium chloride	1.7476
Cesium chloride	1.7627
Zinc blende	1.6381
Wurtzite	1.641
Fluorite	2.519
Rutile	2.408

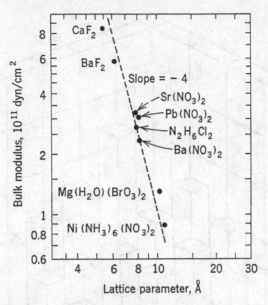

Figure 12.20 Dependence of the bulk moduli of some divalent ionic crystals (fluorite structure) on their crystallographic lattice parameters. Note that even for mixed crystals containing three different ions, the dependence expected from electrostatics is observed.

q is the magnitude of the ionic charge which depends on the valence of a particular ion. The minus sign means that this energy is attractive. The repulsions of the ions are sometimes represented by a power function, and sometimes by an exponential function. Since, over specific ranges of their arguments, these functions can approximate each other, this choice is somewhat arbitrary. Exponential repulsive functions are physically more consistent with quantum mechanics, but power functions are easier to manipulate. Using the latter, the repulsive energy term is:

$$U_r(r) = +A/r^n \tag{12.21}$$

where A and n are constants. By combining Equations (12.19), and (12.20), the following expression for the bulk modulus can be obtained:

$$B = (n - 1)\alpha q^2/18(r_0)^4 \tag{12.22}$$

where r_0 is the equilibrium interionic distance, and n can be empirically determined to be 9–10 (a large exponent is needed if the ion cores are to repulse each other strongly; when n becomes very large the ions respond like rigid spheres). Note the presence of the Keyes parameter (q^2/r_0^4). The inverse fourth power dependence is followed quite well within each homologous series of ionic crystals (Gilman, 1969). Figure 12.20 shows this dependence for a variety of divalent ionic crystals.

The dependence on q is as expected. That is, for divalent ions (such as MgO), the bulk modulus is about four times as large as for monovalent ions of the same size.

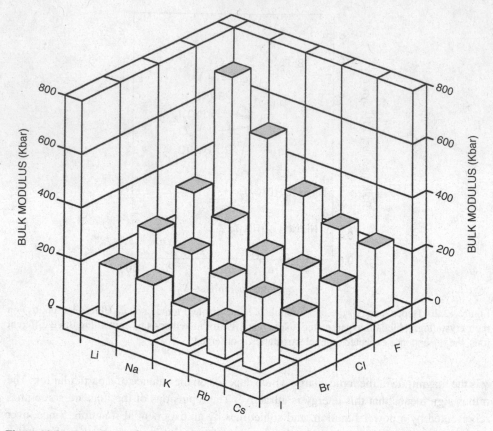

Figure 12.21 Pattern of bulk moduli for the alkali halides. LiF has the maximum bulk modulus, and CsI the minimum.

The full set of alkali halides exhibits a definite pattern of bulk moduli as displayed in Figure 12.21 where the alkali metals are arrayed along one axis, and the halides along another. The largest value of B is for LiF, and the smallest for CsI. The modulus values are for the lowest reported temperatures, mostly 4.2 K. The pattern suggests that there might be a correlation with electronegativity, and indeed there is.

Using Mulliken's definition (electronegativity = [ionization energy − affinity energy]/ $2 = \mu$), and forming the electronegativity difference for each compound, the data for each cation group are plotted (B versus electronegativity difference) in Figure 12.22(a). Each group has its own correlation line. However, if the data are expressed as electronegativity densities (each value in Figure 12.22(a) being divided by the molecular volume), the entire set of data for 20 alkali halides follows a single regression line with $r = 0.97$, see Figure 12.22(b).

A graph similar to Figure 12.22(b) has been made using the average valence electron density as the parameter. The correlation coefficient is similar, indicating that the behavior

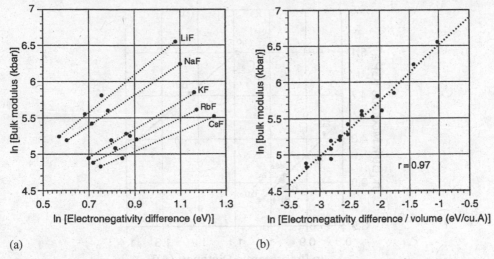

Figure 12.22 Connection between bulk stiffness and electronegativity difference density for the set of 20 alkali halides: (a) bulk modulus versus electronegativity difference (eV); (b) bulk modulus versus electronegativity difference density (eV/Å^3).

is similar to that expressed by Figure 12.4, further indicating that electronegativity is determined by valence electron concentration for these crystals. This suggests that a theory of chemical reactivity might be based on selected mechanical parameters.

12.12 Fluorites

The bulk stiffnesses of the divalent alkaline earth compounds are proportional to the inverse fourth powers of their interionic spacings as Figure 12.20 demonstrates. All of these compounds have the cubic fluorite structure (CaF_2 is the prototype for the series).

12.13 Chalcogenides (oxygen column of the Periodic Table)

The chalcogenide compounds (II–VI) are somewhat unusual in that those containing oxygen behave like ionic compounds, while the others behave like covalent compounds. This is illustrated in Figure 12.23 which shows data for both the divalent oxides and several other chalcogenide crystals. The difference in stability is striking. The divalent oxides follow the inverse fourth power rule, but the other chalcogenides follow a somewhat higher power rule. More striking is that the magnitudes of the stiffnesses for the oxides are about five times greater than those for the other chalcogenides.

The deformation resistances of atomic particles are closely related to their polarizabilities since both involve a change of shape. Figure 12.24 illustrates this for the case of the alkaline oxides where the correlation between the bulk stiffness and the inverse polarizability data of Duffy (1996) is excellent.

Elastic stiffness

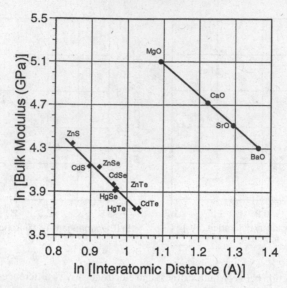

Figure 12.23 Differential behavior within the chalcogenide compounds. The oxides are much stiffer than the other chalcogenide compounds at the same molecular size.

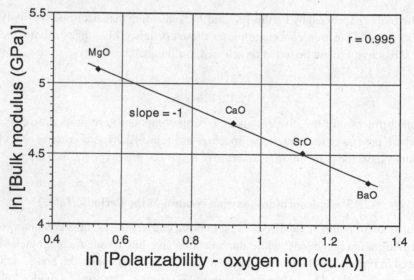

Figure 12.24 Showing that the polarizability (deformability) of the oxygen ion determines the stiffnesses of the alkaline oxides.

12.14 Silicates

As the following table indicates, the bulk moduli of some silicates are similar, suggesting that they are determined principally by the behavior of the silicate groups. The influence of the anion metals is small. The data are from Anderson and Isaak (1995).

Silicate	Bulk modulus (Mbar)
Co_2SiO_4	1.48
Fe_2SiO_4	1.38
Mg_2SiO_4	1.29
Mn_2SiO_4	1.29

12.15 Molecular crystals

The cohesion of molecular crystals through electric dipole–dipole interactions was discussed in Chapter 11. In this chapter the bulk stiffnesses will be considered.

Rewriting Equations (12.2) and (12.4):

$$B = V_0(d^2U/dV^2)_{v=v_0} \tag{12.23}$$

shows the relationship of the bulk modulus to the curvature at the minimum of the binding energy curve at the equilibrium volume V_0. From Chapter 11, Equation (11.30), the Lennard-Jones expression for the binding energy of a face-centered array of polarizable atoms can be written (in terms of the volume, $V = Na^3/4$, of N atoms where a is the lattice parameter, and the nearest-neighbor distance is $r = a/\sqrt{2}$, so $V = Na^3/\sqrt{2}$):

$$U(V) = +A/V^4 - B/V^2 \tag{12.24}$$

with $A = (C/2)N^5\epsilon\sigma^{12}$, and $B = DN^3\epsilon\sigma^6$. In Equation (11.30), the simplification, $C = D = 12$, the sum of the nearest-neighbor pair interactions, was used. Here, the sum over all neighbors is used, so these coefficients become $C = 12.13$ and $D = 14.45$.

At the equilibrium volume, $V = V_0$, $dU/dV = 0$, so:

$$-(4A/V^5) + (2B/V^3) = 0 \tag{12.25}$$

and the solution of this gives a value for the equilibrium volume:

$$V_0 = (2A/B)^{1/2} \tag{12.26}$$

Using this, Equation (12.2) yields the bulk modulus:

$$B = 20A/V_0^5 - 6B/V_0^3 = \alpha(\epsilon/\sigma^3) \tag{12.27}$$

where the coefficient $\alpha = 75.1$, and (ϵ/σ^3) is the Lennard-Jones parameter with ϵ the binding energy and σ the collision diameter. These last two quantities are determined from the van der Waals equation of state for the gas phases, plus the Lennard-Jones equation.

Data for the noble gases are plotted in Figure 12.25. It shows that the stiffnesses are indeed proportional to (ϵ/σ^3) and the value of the proportionality coefficient is about 100,

Elastic stiffness

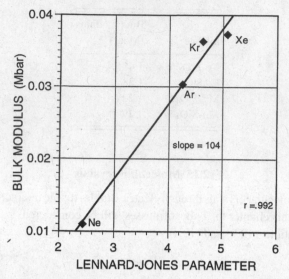

Figure 12.25 Comparison of measurements and theory based on the Lennard-Jones energy relation for "molecular" crystals. The inert gas crystals in this case are taken to be prototypes.

so use of the Lennard-Jones potential is approximately validated. Also, it shows that the same forces that cause non-ideality in gases (that is, deviations from $PV = nRT$) cause the elastic stiffness of noble gas crystals, and of other molecular crystals.

12.15.1 Note added in proof

The bulk modulus of Os has recently been remeasured. A value of 4.62 Mbar was determined (Cynn, H., Klepis, J.E., Yoo, C., and Young, D.A. (2002). *Phys. Rev. Lett.*, **88** (13), 135701-1). This is 4.3% larger than the value for diamond (4.43 Mbar). Thus osmium is stiffer than diamond (in hydrostatic compression), albeit diamond is still the champion for hardness.

References

Anderson, O.L. and Isaak, D.G. (1995). Elastic constants of mantle materials at high temperatures. *Mineral Physics and Crystallography – A Handbook of Physical Constants*, AGU Reference Shelf 2. Washington, DC: American Geophysical Union.

Aryasetiawan, F. and Karlsson, K. (1994). *Phys. Rev. Lett.*, **12**, 1679.

Atkins, P.W. (1991). *Quanta*, p. 151. Oxford: Oxford University Press.

Banerjea, A. and Smith, J.R. (1988). Origins of the universal binding energy relation, *Phys. Rev. B*, **37**, 6632.

Burdett, J.K. (1995). *Chemical Bonding in Solids*. New York: Oxford University Press.

Chen, C.J. (1991). Attractive interatomic force as a tunneling phenomenon, *J. Phys.: Condens. Matter*, **3**, 1227.

Cottrell, A.H. (1988). *Introduction to the Modern Theory of Metals*. London: The Institute of Metals.

Cottrell, A.H. (1997). *The Art of Simplification in Materials Science*, MRS Bulletin, Volume 22 (5), p. 15. Pittsburgh, PA: Materials Research Society.

Duffy, J.A. (1996). Optical basicity: a practical acid-base theory for oxides and oxyanions, *J. Chem. Educ.*, **73** (12), 1138.

Edwards, P.P. and Sienko, M.J. (1983). What is a metal?, *Int. Rev. Phys. Chem.*, **3**, 83.

Gibbons, P.C., Schnatterly, S.E., Risko, J.J., and Fields, J.M. (1976). *Phys. Rev. B*, **13**, 2451.

Gilman, J.J. and Roberts, B.W. (1961). Elastic constants of TiC and TiB_2, *J. Appl. Phys.*, **32** (7), 1405.

Gilman, J.J. (1966). Monocrystals in mechanical technology, Campbell Memorial Lecture, *Am. Soc. Met. Trans. Q.*, **59** (4), 597.

Gilman, J.J. (1969). *Micromechanics of Flow in Solids*, Chapter 2. New York: McGraw-Hill.

Gilman, J.J. (1971). Bulk stiffnesses of metals, *Mater. Sci. Eng.*, **7**, 357.

Gilman, J.J., Cunningham, B.J., and Holt, A.C. (1990). Method for monitoring the mechanical state of a material, *Mater. Sci. Eng.*, **A125**, 39.

Gilman, J.J. (1997). Chemical and physical hardness, *Mater. Res. Innovat.*, **1**, 71.

Gilman, J.J. (1999). Plasmons at shock fronts, *Philos. Mag. B*, **79**, 643.

Gschneidner, K.A. and Eyring, L. (Editors) (1978). *Handbook on the Physics and Chemistry of Rare Earths*, Volume 1, *Metals*, p. 591. Amsterdam: North-Holland.

Harrison, W.A. (1980). *Electronic Structure and the Properties of Solids*. San Francisco, CA: W.H. Freeman.

Keyes, R.W. (1962). Elastic properties of diamond-type semiconductors, *J. Appl. Phys.*, **33**, 3371.

Kittel, C. (1996). *Introduction to Solid State Physics*, 7th edn., p. 59. New York: Wiley; see also 5th edn., 1953.

Ledbetter, H. (1983). Elastic properties. In *Materials at Low Temperatures*, ed. R.P. Reed and A.F. Clark, Chapter 1. Metals Park, OH: American Society for Metals.

McWeeny, R. (1990). *Coulson's Valence*, 3rd edn. Oxford: Oxford University Press.

Moruzzi, V.L. and Marcus, P.M. (1993). Trends in bulk moduli from first-principles total-energy calculations, *Phys. Rev.*, **48**, 7665.

Nakamura, M. (1995). In *Intermetallic Compounds*, ed. J.H. Westbrook and R.L. Fleischer, Volume 1, Chapter 37, Elastic properties, p. 873. New York: Wiley.

Parr, R.G. and Yang, W. (1989). *Density-Functional Theory of Atoms and Molecules*, p. 95ff. New York: Oxford University Press.

Pauling, L. (1960). *The Nature of the Chemical Bond*, 3rd edn. Oxford: Oxford University Press.

Pearson, R.G. (1997). *Chemical Hardness*. New York: Wiley-VCH.

Pettifor, D. (1995). *Bonding and Structure of Molecules and Solids*. Oxford: Clarendon Press.

Pines, D. (1963). *Elementary Excitations in Solids*. New York: W.A. Benjamin.

Pugh, S.F. (1954). Relations between the elastic moduli and the plastic properties of polycrystalline pure metals, *Philos. Mag.*, **45**, 823.

Rose, J.H., Ferrante, J., and Smith, J.R. (1981). Universal binding energy curves for metals and bimetallic interfaces, *Phys. Rev. Lett.*, **47**, 675.

Smith, N.V. (1970). *Phys. Rev. B*, **2**, 2840.

Sutton, A.P. (1994). *Electronic Structure of Materials*. Oxford: Clarendon Press.

Whang, U.S., Arakawa, E.T., and Callcott, T.A. (1972). *Phys. Rev. B*, **6**, 2109.

Wood, R.W. (1933). *Phys. Rev.*, **44**, 353.

13

Shear modulus

13.1 General comments

Many orders of magnitude lie between the elastic response of a soft rubber (elastomer) and that of rigid diamond, more than six orders, in fact. The question for physical chemistry is: what determines the shear stiffness of a given substance? This has some sub-divisions. The two main ones are: *enthalpic* and *entropic* elasticity. This chapter discusses the first of these, and Chapter 14 discusses the latter.

The distinction between shear and dilatation is illustrated in Figure 13.1 which shows an undeformed circle at 13.1(a), the effect of shear on the circle at 13.1(b), and the effect of dilatation at 13.1(c). Note that shear does not change the area enclosed by the circle (for small deformations).

Shear moduli measure resistance to shape changes. For small shears, the volume is constant, so these moduli are conjugate to the bulk moduli which measure volume changes at constant shape.

As mentioned in Chapter 4, it was not clear until about 1887 whether one or two independent constants are needed to describe the shear responses of isotropic elastic materials (and 15 or 21 to describe trigonal crystals) (Timoshenko, 1983). In 1821, Claude Navier had proposed an atomic theory in which only central forces acted between the atoms. This related the shear and bulk moduli to one another, leaving just one independent constant for the isotropic case. Others, e.g., Stokes and Poncelet, thought that the resistance to shear was fundamentally different from the resistance to compression, so two constants were needed. After some 65 years had passed, the controversy was settled when Waldemar Voigt made careful measurements of the elastic constants of monocrystals, and found that, in general, three independent constants are needed for cubic crystals, and two for isotropic solids.

The most symmetric crystals are cubic ones. Imagine eight balls connected by twelve springs to form a cube with a spring along each edge. If such a cube is stretched along the body diagonal that passes through two of the diagonally opposite corner balls, the response is quite different from the response that occurs if the cube is stretched along a diagonal of one of the faces. Thus two independent shear constants, plus one bulk constant, are needed to describe the elastic state of a deformed cubic crystal.

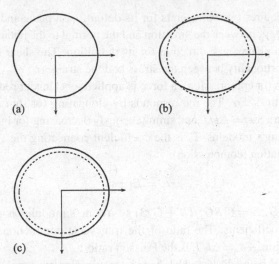

Figure 13.1 Schematic illustration of pure dilatational deformation (c), and pure shear deformation (b). (a) represents the undeformed body.

Any two independent shear constants can be used, but it is convenient to use a standardized set. The set chosen by convention can be understood as follows. Imagine a square piece of cloth having a simple basket weave. Let the threads run parallel to the sides of the square (the warp parallel to two opposite sides, the woof parallel to the perpendicular opposite sides). Pull the cloth along one of the diagonals of the square, and push equally along the other diagonal. The cloth will be quite compliant. The square shape will become sheared into a diamond-like shape. The pertinent elastic constant is called C_{44}. Now pull the cloth in a direction parallel to two of its edges, and push equally along the perpendicular direction. The cloth will be stiff because it is being pulled parallel to one set of its threads. The overall shape will again be sheared, and the elastic constant that applies is called C'. It is easy to see why this is a good choice for the two shear constants.

13.2 Shear stiffnesses

The most simple way to change the shape of a homogeneous sphere of material is to stretch (or compress) it along one axis, while allowing one of the perpendicular axes to shorten (or lengthen), holding the length of the other perpendicular axis fixed. This keeps the volume constant (for small deformations), thereby causing a pure shape change. The initial spherical shape becomes an ellipsoid with one axis longer, one axis shorter, and one axis unchanged relative to the radius of the sphere. The spherical symmetry becomes reduced to tetrahedral, or rhombohedral, symmetry.

Pure shear is not easily achieved in practice. For quasi-static loading, the only available method is to apply torsion to a thin-walled tube. For dynamic loading, transducers designed for shear deformation can produce shear waves that remain "pure" for a short time.

A shear strain requires two parameters for its definition, a plane and a direction. These define a shear angle, γ, between the direction and the normal to the plane. Similarly, a shear stress, τ, requires a plane and a direction for its definition. The shear modulus, G, is the coefficient of proportionality between the stress and the strain.

In simple tension (or compression), a force is applied to a slender rod parallel to its axis creating a tensile stress $\pm\sigma$. The rod responds by elongating (or shortening) a fractional amount (axial strain), $\pm\epsilon = \Delta l/l$, and simultaneously decreasing (or increasing) its cross-sectional area. Young's modulus, E, is the coefficient connecting the applied tension and the fractional elongation (compression):

$$\pm\sigma = E(\pm\epsilon) \qquad (13.1)$$

In terms of B and G, $E = (3BG)/(B + G/3)$ so Young's modulus is a composite of the two fundamental coefficients. The ratio of the transverse contraction, $-(\Delta d/d)$, to the longitudinal extension, $+\epsilon = \Delta l/l$, is the Poisson ratio ν.

So far, the material being discussed has been isotropic. However, real materials consist of arrays of atoms or molecules in the form of crystals which are not isotropic, except for special cases. For crystals, Equation (13.1) becomes considerably more complex (Chapter 4). The stress, and the strain, become second-order tensors σ_{ij} and ϵ_{kl}, where one of the indices designates the plane across which the stress (or strain) acts, while the other index designates a direction either parallel or perpendicular to the plane. For extensions, $k = l$, while for shears and rotations $k \neq l$. The connecting coefficients are elastic stiffnesses C_{ijkl} and Equation (13.1) becomes:

$$\sigma_{ij} = C_{ijkl}\epsilon_{kl} \qquad (13.2)$$

The crystals of most interest have cubic symmetry which drastically reduces the necessary stiffness coefficients from 21 to 3.

Since only three coefficients are needed to describe cubic crystals, four indices are not needed. Therefore, by convention, only two indices are used; and the three coefficients are C_{11}, C_{12}, and C_{44}. Then the elasticity matrix (with the 1, 2, 3 crystal axes taken parallel to the $1, 2, 3$ coordinate axes) is:

	ϵ_{11}	ϵ_{22}	ϵ_{33}	ϵ_{23}	ϵ_{31}	ϵ_{12}
σ_{11}	C_{11}	C_{12}	C_{12}	0	0	0
σ_{22}	C_{12}	C_{11}	C_{12}	0	0	0
σ_{33}	C_{12}	C_{12}	C_{11}	0	0	0
σ_{23}	0	0	0	C_{44}	0	0
σ_{31}	0	0	0	0	C_{44}	0
σ_{12}	0	0	0	0	0	C_{44}

$$(13.3)$$

where the C_{11} refer to deformations in the same direction as the stresses; the C_{44} refer to shear deformations; and the C_{12} give the deformations transverse to the principal stresses.

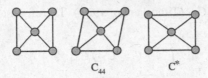

Figure 13.2 Demonstrating the meanings of the two shear coefficients, C_{44} and C^*. At the left is a sketch of the undeformed square (cube). The angle between the diagonal lines measures the magnitude of the shear strain.

For a change of volume, V, $\Delta V/V$ is the average of the three axial strains, $\Delta V/V = (\epsilon_{11} + \epsilon_{22} + \epsilon_{33})/3$. The bulk modulus B relates this to applied hydrostatic pressure P for which the three principal stresses are equal, so $\sigma_{11} = \sigma_{22} = \sigma_{33} = P$. Thus:

$$B = (C_{11} + 2C_{12})/3 \tag{13.4}$$

The single shear modulus of an isotropic solid becomes two moduli for a cubic crystal. Shear has two-fold symmetry so it is expected that the two moduli will refer to planes of two-fold symmetry. Of the three principal planes of cubic crystals, {100}, {110}, and {111}, the first has four-fold symmetry, the second two-fold symmetry, while the third has three-fold symmetry. Thus the two shear stiffnesses must refer to the first two planes. From Equation (13.3) and the orientations of the axes, it can be seen that C_{44} refers to strains in the [010] directions (cube edges) on the corresponding {100} planes (cube faces). The deformation is that which occurs to a circle lying on a (100) plane and being stretched (or compressed) along one of the cubic diagonal axes.

The other shear mode corresponds to the same circle being stretched (or compressed) along one of the cube edge axes. Thus one of the modes can be transformed into the other by a rotation of 45°. The coefficient of the second shear mode will be called C^*, and it equals $(C_{11} - C_{12})/2$ (Figure 13.2).

The condition for isotropy is that there be only one shear constant, so that $C_{44} = C^*$. From these definitions, inspection of Figure 13.2 indicates that if the interiors of the boxes are uniform, the elastic response will be isotropic. Only if some structure exists within the boxes will the responses to the two types of shear be different. Even if there is structure, the response *may be* isotropic but usually will not be.

The deviation from the isotropy condition can be described in terms of the ratio of the two shear constants, known as Zener's anisotropy constant A:

$$A = C_{44}/C^* = 2C_{44}/(C_{11} - C_{12}) \tag{13.5}$$

This equals unity when a cubic crystal is isotropic.

In the history of elasticity theory, a controversy raged for nearly 100 years as to whether $A = 1$ always (the rari-constant theory), or whether sometimes $A \neq 1$ (the multi-constant theory). The controversy was finally resolved experimentally by Voigt in favor of the multi-constant theory (Timoshenko, 1983). During the controversy, Cauchy showed that if solids consist of atoms, and the only forces between the atoms exist along the lines connecting their centers, then $C_{44} = C_{12}$ (Timoshenko, 1983). Therefore, another ratio of interest is the Cauchy constant:

$$\Gamma = C_{12}/C_{44} \tag{13.6}$$

Deviations of Γ from unity measure the magnitudes of the bending force constants of the chemical bonds in the structure relative to their stretching constants.

Still another ratio that is sometimes useful is the ratio of the minimum shear modulus G_{min} to the bulk modulus. This might be called the "solidity index", S:

$$S \equiv 3/\pi(G_{min}/B) \tag{13.7}$$

S varies from values near zero for very soft solids to about unity for diamond. The smaller S is, the more liquid-like the material; and the larger it is, the greater the relative rigidity (stability) of the material. Thus it is one of the factors involved in ductility (Cottrell, 1997). For small S, materials tend to be ductile, while for large S, they tend to be brittle. For cubic materials, $G_{min} = C_{44}$, or $C^\star$, depending on the crystal structure.

For less symmetric crystals more constants are needed, up to 21 for crystals with trigonal symmetry (Chapter 4). Only the cubic case will be considered here. The first constant for cubic crystals is C_{11}. When a strain is applied along one of the three edges of the symmetry cube, and is multiplied by C_{11}, the normal stress in that direction is obtained. The simplest way of doing this is to apply a plane, uniaxial elastic wave by means of a thin transducer. There are three equivalent cases, one for each cube axis, so there are three constants; but because of the symmetry they equal one another numerically.

Similarly, a strain applied along one of the cube edges when multiplied by C_{12} gives the stresses parallel to the two edges perpendicular to the strained one. By symmetry, there are three equivalent directions for the applied strains, and two perpendicular stresses for each of them, so six numerically equal constants describe the corresponding stresses.

Finally, if a shear strain is applied with the shear plane parallel to one of the cube faces, then C_{44} multiplied by the strain gives the shear stress, and there are three equivalent faces described by means of three equivalent constants.

The relationships between the shear moduli and the atomic properties are complex because, unlike the scalar bulk modulus, the shear moduli are tensors, for which both planes, directions, and crystallographic orientations must be specified.

13.3 The Cauchy relations

In the early history of the subject (Love, 1944), much of which was developed by Navier and Cauchy, there was interest in the connections between the behavior of "molecules" and the macroscopic stresses and strains (many people did not believe in "atoms" at the time). Cauchy showed that if a crystal structure has a center of symmetry, and the forces of attraction and repulsion act along lines connecting the centers of the "molecules", then only two constants are needed to describe the elastic stiffness of a cubic crystal because $C_{12} = C_{44}$.

Table 13.1 *Elastic constants of some*
isoelectronic alkali halides

Halide	C_{11}	C_{12}	C_{44}
LiF	1.24	0.43	0.65
NaCl	0.58	0.12	0.13
KBr	0.35	0.056	0.051
RbI	0.32	0.036	0.029

Subsequently it has been found that most solids cannot be described by means of forces that act only along the axes between their atoms. In fact, although there have been several attempts to use just two force constants at the atomic level to describe the macroscopic elastic stiffnesses (one radial plus one angular, or one radial between nearest-neighbor atoms plus another radial between next nearest-neighbor atoms), it seems likely that, in general, at least three microscopic constants are needed (one radial plus two angular) (Harrison, 1980; see also, Phillips, 1973).

A simplified version of the Cauchy theory was described in Chapter 4. In that version, springs with two different constants, k_1 and k_2, were used to create interaction forces between "atoms" (Figure 4.1). The crystals that most nearly match the assumptions of the spring constant model are the alkali halides. They consist of alternating negative and positive ions in a face-centered cubic array (Figure 12.18(a)). Table 13.1 lists some elastic constants for a few isoelectronic alkali halides. For them, it may be seen that $C_{12} \approx C_{44}$. The data are for the lowest temperature available (Simmons and Wang, 1971).

13.4 Simple metals

There are at least three simple analytic ways to consider the effect of shear on the energy of a metal, and thereby arrive at a value for the shear modulus. They are all based on electrostatics; and they are all vectorial rather than tensorial so their usefulness is limited, but they do indicate the origin of the shear stiffness. The methods are:

(1) effective atom (Gilman, 1966),
(2) dipole polarizability (Gilman, 1997),
(3) quadrupolar polarizability (Machlin and Whang, 1978).

13.4.1 Effective atom method

For an isotropic metal we can make a simple model of the shear stiffness based on an equivalent atom. Keep in mind that the model is a severe simplification.

All of the atoms in a crystal are the same so we dissect one of them out and take it to be representative. As we did for the bulk modulus, we start with a positive nucleus immersed in a uniform electron cloud of mean square radius, R. The model is feasible because a shear distortion (which converts the sphere into an ellipsoid) has no first-order effect on either the kinetic energy of the electron or the electron–electron interaction. That leaves the ion–cloud interaction, $-3q^2/2r$.

A shear strain, γ, converts a sphere into an ellipsoid with the three axes:

$$a = R$$
$$b = R(1 - \gamma/2)$$
$$c = R(1 + \gamma/2)$$

And, for small strains, the electrostatic energy of the ellipsoid is:

$$\phi = 3q^2/2(abc)^{1/3} \tag{13.8}$$

During shear, the volume, $4\pi r^3/3$, remains unchanged, so the electrostatic energy per unit volume becomes:

$$\Phi(\gamma) = (9q^2/8\pi R^4)[1 - \gamma^2/4]^{1/3} \tag{13.9}$$

To obtain the shear modulus, G, we determine the second derivative, $\Phi''(\gamma)$, neglecting the γ^2 term:

$$G = 3q^2/32\pi R^4 \tag{13.10}$$

For example, in aluminum metal the atomic radius is 1.25 Å, so Equation (13.10) yields $G = 28\,\text{GPa}$. This compares well with the observed value, 23 GPa. Also, for Al, the constant in Equation (13.10) is 0.27, and therefore closer to the experimental value than the value of 0.28 given by the detailed theory of Fuchs (1935). In view of the simplicity of the model, this should only be taken to mean that the shear modulus has an electrostatic origin just as does the bulk modulus.

Real metals cannot be represented by equivalent spheres. However, they might be represented by pairwise potentials, that is, by interactions that depend only on the distances between atoms, and not upon angles between three, or more, atoms. This assumption leads, for isotropic symmetry, to a specific ratio between the shear and bulk moduli, namely:

$$G/B = 3/5 \tag{13.11}$$

The ratio of Equation (13.10) to Equation (12.8) is 0.63 ($\approx 3/5$) so the simple models presented here for G and B are consistent with pairwise models.

13.4.2 Dipole polarizability

The second method takes advantage of the equivalence of electric and stress fields. Thus shear strains and electric polarization have similar effects on the energy of a solid. If complete polarizability tensors were known this method could yield the shear compliance tensor, but at present it is limited to the isotropic case where there is only one value of the polarizability.

In an electric field, E_j, an atom develops a dipole moment, $\mu_i = \alpha_{ij} E_j$, which represents a separation of the centers of positive and negative charge. That is: $\mu_i = q x_i$ where q is the electron charge for a monovalent atom, x_i is the separation distance, and α_{ij} is the polarizability tensor. This elongates the atom a small amount converting it into an ellipsoid with major axis $R(1 + x_i)$ and, at constant volume, minor axes $R(1 - x_i/2)$. Therefore, the equivalent shear strain is $\gamma = x/R$.

To obtain an expression for the shear modulus we start by equating the electrical and mechanical forces that resist the shape change described by γ. The electrical force is Eq, and the mechanical force is the gradient of the strain energy, or $(4\pi/3)GRx$. Combining these equations and solving for G yields:

$$G = (3/4\pi)(q^2/\alpha R) \tag{13.12}$$

so, the shear modulus G is inversely proportional to the polarizability. Furthermore, since $\alpha \approx (4\pi/3)R^3$ the shear modulus may take the form:

$$G = (q^2/R^4)/30 \tag{13.13}$$

which displays the Keyes parameter, q^2/R^4, as might be expected.

This expression yields $G = 0.07$ Mbar for sodium which compares well with the experimental value of 0.05 Mbar.

13.4.3 Quadrupole polarizability

The third method, due to Machlin and Whang (1978) is based on the idea that atomic shape changes are reflected in the coupling of the atom's electrons to the quadrupole moment of its nucleus. A measure of this is the quadrupolar polarizability α_4 (Blatt and Weisskopf, 1991):

$$\alpha_4 = -qQ/(\partial E/\partial z) \tag{13.14}$$

where Q is the nuclear quadrupole moment, and $(\partial E/\partial z)$ is the electric field gradient of the nucleus. As before, q is the electronic charge.

Consider a neutral sphere, sheared to form an ellipsoid with axes, a, a, and c. Let the rms radius be $R^2 = (a^2 + c^2)/2$. Then the strain is $(c^2 - a^2)/2R^2 = \gamma$; so the quadrupole moment of the strained nucleus is $Q = -(4\epsilon q R^2)/5$.

The electric stress, $\Sigma = (1/R)(\partial E/\partial z) = (qQ\gamma)/(R\alpha_4)$, and the corresponding mechanical stress $\sigma = C_{44}\gamma$. Equating the stresses and solving:

$$C_{44} = qQ/R\alpha_4 \tag{13.15}$$

Since $Q \sim R^2$ and $\alpha_4 \sim R^5$, $C_{44} \sim q^2/R^4$ (Keyes parameter). Thus, whether monopoles, dipoles, or quadrupoles are used, the forces are electrostatic. For a more detailed discussion of the behavior described by Equation (13.15), see Machlin and Whang (1978). Figure 13.3 illustrates their data, demonstrating that the electrons in atoms are indeed coupled to their

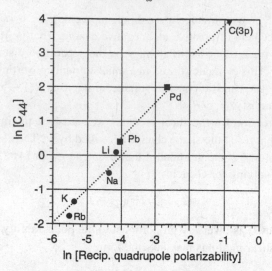

Figure 13.3 Dependence of the shear moduli, C_{44}, on the reciprocal nuclear quadrupole-polarizabilities (ln–ln plot). Data from Machlin and Whang (1978).

nuclei. Although there are too few data to be sure, it appears that there is some effect of crystal structure as well as the overall electron density because the dotted lines in the figure are not superimposed.

Since the quadrupole polarizability is proportional to the dipole polarizability (Mahan and Subbaswamy, 1990), and the latter is by far the easier to measure, the correlation of the shear stiffness with quadrupole polarizability has minimal practical interest, but it is interesting that the coupling is affected by shear strain.

13.5 Failure of radial potentials

For many real materials, there are large deviations from the idealized picture in which only radial forces act between the atoms. The histogram of Figure 13.4 demonstrates this. It includes all of the available data for the solid elements (Gilman, Cunningham, and Holt, 1990), and it shows that, instead of being constant, the ratio G/B varies markedly with position in the Periodic Table of the Elements. The range of the ratio is a factor of 14. Soft, ductile elements lie toward the left of the histogram, while hard brittle ones lie toward the right. The ratio at the mode is 0.45, which is quite different from the 0.6 expected from central force models. Thus there are large chemical effects that influence the elastic moduli. Metals cannot be viewed simply as arrays of spheres. This becomes even more apparent if we consider the elastic anisotropies of metals.

Let us look somewhat further at the apparent connection between G/B and ductility. The pure f.c.c. metals are conventionally considered to be ductile, but Table 13.2 indicates that there is a systematic decline in ductility as G/B increases by a factor of about four in going from Pb to Ir, see also Cottrell (1997). This list could probably be improved by the use of low temperature data.

Table 13.2 *G/B ratios for f.c.c. elements*

Metal	G/B
Pb	0.14
Au	0.16
Pt	0.22
Pd	0.23
Ag	0.29
Au	0.35
Al	0.37
Ni	0.41
Ca	0.43
Sr	0.51
Rh	0.55
Ir	0.57
Th	0.58

Moduli ratio ranges

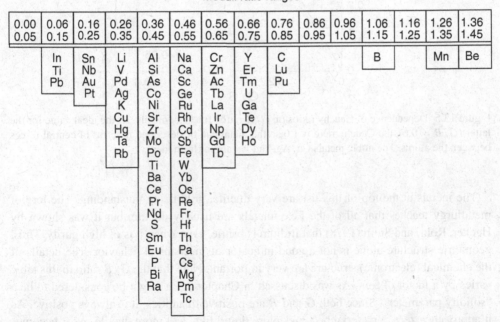

0.00 0.05	0.06 0.15	0.16 0.25	0.26 0.35	0.36 0.45	0.46 0.55	0.56 0.65	0.66 0.75	0.76 0.85	0.86 0.95	0.96 1.05	1.06 1.15	1.16 1.25	1.26 1.35	1.36 1.45
	In	Sn	Li	Al	Na	Cr	Y	C			B		Mn	Be
	Ti	Nb	V	Si	Ca	Zn	Er	Lu						
	Pb	Au	Pd	As	Sc	Ac	Tm	Pu						
		Pt	Ag	Co	Ge	Tb	U							
			K	Ni	Ru	La	Ga							
			Cu	Sr	Rh	Ir	Te							
			Hg	Zr	Cd	Np	Dy							
			Ta	Mo	Sb	Gd	Ho							
			Rb	Po	Fe	Tb								
				Ti	W									
				Ba	Yb									
				Ce	Os									
				Pr	Re									
				Nd	Fr									
				S	Hf									
				Sm	Th									
				Eu	Pa									
				P	Cs									
				Bi	Mg									
				Ra	Pm									
					Tc									

Figure 13.4 Histogram demonstrating the non-ideality of the elastic behaviors of most of the chemical elements. The frequencies of the *G/B* ratios are plotted. The ideal value is 0.6, but the real values range from 0.1 to 1.4. Ductility increases to the left.

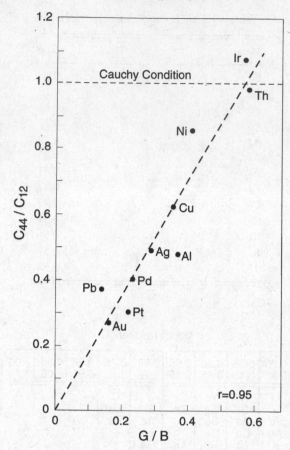

Figure 13.5 Dependence of Cauchy ratios on G/B ratios for f.c.c. metals. At the ideal value for the latter ($G/B = 0.6$), the Cauchy ratio is 1.0 so the metal can be described in terms of central forces between the atoms. The noble metals Cu, Ag, and Au are far from ideal.

The metals at the top of this list are very ductile, gold being outstanding. The lore of metallurgy teaches that all of the f.c.c. metals are usually ductile, but it was shown by Hecker, Rohr, and Stein (1978) that iridium is brittle, even when it is of high purity. Thus, geometric structure alone is not a good indicator of mechanical behavior. The details of the chemical (electronic) structure are very important. Note that the G/B ratio in this table varies by a factor of four. As was discussed in Chapter 12, it might be considered to be a "solidity parameter". Since both G and B are positive quantities, it is always positive. As it approaches zero, a material becomes more liquid like, and more ductile. As it becomes large, a material becomes increasingly rigid, and more brittle. For real materials it does not exceed about 1.3 (diamond).

In spite of their simple crystal structures, largely based on the close packing of spherical atoms, metals usually do not satisfy either the central force rule ($C_{12} = C_{44}$), or the isotropy rule ($A = 1$). This confirms that a great deal of chemistry is involved in their cohesion as emphasized long ago by Pauling (1960), and developed by Harrison (1980), Cottrell (1988), and Pettifor (1995). This complex subject will not be surveyed here.

Figure 13.5 shows that there is a connection between G/B and the Cauchy ratio C_{44}/C_{12} for the f.c.c. elements. As G/B increases, C_{44}/C_{12} increases with a large amount of correlation (coefficient 0.95). Notice that the brittle metal, Ir, has a Cauchy ratio close to the ideal, while the metal that may have the most ductility, namely Au, lies far from it.

Overall, it is apparent that although the atoms in these f.c.c. metals have a simple structural arrangement (close-packing), the distributions of electrons in them are not so simple.

13.6 Alkali metals

The alkali metals are of special interest because the basis of their structures (predominantly body-centered cubic) (b.c.c.) is not immediately obvious. As described in Chapter 12, their ion cores (single ionization) are relatively small compared with their neutral atoms (Table 12.1), and their electrical properties are consistent with their valence electrons being nearly free. Thus they consist of small ion cores immersed in an electron liquid. Therefore, it might be expected that their atoms would have spherical symmetry, and that they would exhibit atomic close-packing (either f.c.c. or h.c.p.). For spheres of a given size, the spacial packing fraction is 0.74 for the f.c.c. structure, and 0.68 for the b.c.c. structure. However, Frank (1992) has pointed out that metals need not be considered to consist of hard spheres; they may also be considered to consist of assemblies of nuclei and electrons where the nuclei are concentrated at approximate points, while the electrons are distributed approximately continuously throughout each atomic volume.

Given a uniform distribution of charge, the cohesion of an atom is determined by the electrostatic energy of interaction between its nucleus and its electrons. Since electrostatic energy depends inversely on the distance between charged particles, an important factor for the energy of an atom is the closeness of its nucleus to any point in an atomic subcell of its crystal. The subcells are Wigner–Seitz cells (also known as Voronoi cells). They are formed by connecting a given nucleus to its nearest neighbors by straight lines, and then bisecting each line by a perpendicular plane. This forms a polyhedron enclosing each nucleus. These polyhedra are convex and space filling.

Let a_f be the f.c.c. lattice parameter, and a_b the b.c.c. lattice parameter. Then, the volume per atom in the f.c.c. structure is $v_f^a = a_f^3/4$, and in the b.c.c. structure it is $v_b^a = a_b^3/2$. Therefore, the condition for constant atomic volume is $v_f^a = v_b^a$, or $a_f = 1.26a_b$. Let the distance between nearest neighbours be b_i. This distance in the f.c.c. structure is $b_f = a_f/\sqrt{2}$; and in the b.c.c. structure it is $b_b = (\sqrt{3}/2)a_b$. Therefore, the ratio is $b_b/b_f = \sqrt{1.5}/1.26 = 0.972$; and the nearest neighbors in the b.c.c. structure are about 3% closer electrically than in the f.c.c. structure at constant atomic volume. Thus, the attractive Coulomb potential favors the b.c.c. structure by a small amount, while the kinetic energy of a valence electron is the same for both structures since the atomic volumes are the same.

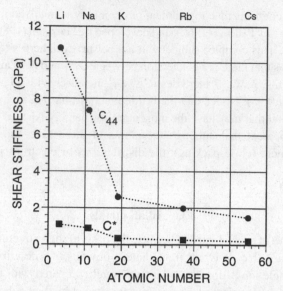

Figure 13.6 Shear stiffnesses of the alkali metals emphasizing the large differences between the C_{44} and C^* values for Li and Na.

Although the nominal coordination number is smaller for the b.c.c. structure than for the f.c.c. structure (8 versus 12), the "effective coordination" number is 14 (8 + 6), since the six next-nearest neighbors are only 14% further apart than the nearest neighbors.

Note that, in disagreement with conventional wisdom, there is no a priori reason to attribute directionality to the bonding in the b.c.c. structure.

The above paragraphs are consistent with the importance of electron density in determining stability. The sizes of the atoms are set by the de Broglie and Pauli Principles, then the valence electron concentrations determine several of the other properties. In addition to total cohesive energy, valence electron density determines the energies of fluctuations, that is, plasmon energies. For short distances between them, plasmon oscillations interact strongly. These interactions will be discussed in more detail in Chapter 20 on surface energies. They may also have small effects on the stabilities of metallic structures.

Figure 13.6 shows how the two shear moduli, C_{44} and $C^* = (C_{11} - C_{12})/2$, vary with atomic number for the alkali metals (plotted according to their atomic numbers). In the case of Li, C_{44} is almost an order of magnitude greater than C^*. This may be a result of the fact that the atomic shear planes are about 40% more widely separated for C^* than for C_{44} in the b.c.c. structure. The anisotropy coefficients for the alkalis are displayed in Figure 13.7. They are all very large, being of the order of what might be expected if the difference in stiffness is attributable to a difference in dispersion forces (plasmon–plasmon interactions) across the shear planes. If the dispersion forces vary as $1/r^6$ where r is the distance between the shear planes, then the anisotropy coefficient would be expected to be $(2^{1/2})^6 = 8$ which is consistent with Figure 13.7.

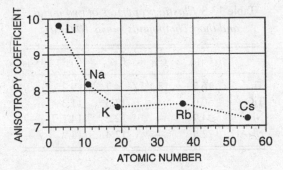

Figure 13.7 Anisotropy coefficients of the alkali metals. Note the extreme value of nearly 10 for Li.

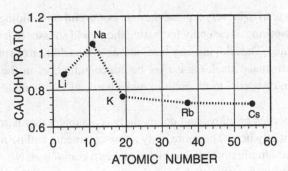

Figure 13.8 Cauchy ratios for the alkali metals. They are near the ideals in contrast to the extreme anisotropy coefficients of Figure 13.7. Na is bound by central radial forces.

The Cauchy ratios for the alkali metals are shown in Figure 13.8. Except for sodium, they are neither close to, nor far from, unity. Since the anisotropy coefficient is eight for sodium, pairwise forces do not act for sodium, so the Cauchy rule is not symmetric. That is, if only central, pairwise, forces act, then the Cauchy ratio must be unity; but if the ratio is unity there may, or may not, be only central, pairwise, forces.

13.7 Compounds

The most simple compounds are the alkali halides. However, the most important compounds for use in structural applications are:

(1) the intermetallic compounds,
(2) the "hard metals" consisting of transition metals combined with one, or more, of the metalloids (boron, carbon, or nitrogen).

For applications in solid-state electronics, the III–V, II–VI, and I–VII compounds are also important. In addition, various oxides and other chalcogenides play important roles for optical materials, and for high temperature materials as well as other ceramics. Silicates form most of the rocks of the Earth so they are important in geophysics.

Table 13.3 *Elastic properties of two metals
and their compounds (units 10 GPa)*

	C_{11}	C_{33}	C_{44}	C_{12}	C_{13}
Ti	16.2	18.1	4.7	9.2	6.9
TiC	50.0	—	17.5	11.3	—
Ni	24.7	—	12.5	14.7	—
Ni$_3$Al	22.4	—	12.4	14.9	—

13.7.1 Alloys and intermetallic compounds

High purity metals tend to be very soft, so they are not useful for building strong structures. Alloying, or cold-working, is essential for raising their yield stresses to levels that are useful for building structures. The alloying may be dilute, as in the addition of small concentrations of carbon to iron to make steel. Or it may be concentrated, as in the addition of large amounts of carbon, or boron, to titanium to make hard, stiff compounds such as TiC and TiB$_2$.

The elastic properties of alloys and compounds is a subject that is too extensive to be considered substantially here. Therefore, only a few comments will be made.

Large amounts of aluminum added to nickel makes a compound, Ni$_3$Al, with nearly the same elastic constants as pure nickel, but very much greater hardness, especially at high temperatures where this compound is the basis of the best high performance alloys for aircraft engines. Table 13.3 shows some data for these examples.

The middle column of the table, for C_{44}, is the most important for our purposes. In the Ti case, the metalloid carbons, C, fit into the interstices between planar, hexagonal close-packed arrays of Ti atoms. Half of the interstitial holes are occupied. This makes the shear stiffness of the compound 3.7 times greater than that of the pure metal through strong metal–metalloid interactions, and an increase in the valence electron density.

In the case of Ni$_3$Al, the Al substitutes for 25% of the Ni atoms in the f.c.c. structure. This has little effect on the shear stiffnesses, but a very large effect on the hardness. Although hardness and stiffness usually correlate with each other, this is an example where the hardness is determined by an aspect of the crystal structure that affects dislocation mobility (rather than the elastic behavior). Since dislocation mobility in Ni$_3$Al decreases with increasing temperature up to rather high temperatures, an entropic factor appears to play an important role.

As another example, compare silicon and copper. They have nearly the same shear moduli but the Vickers hardness of pure Si is 50 times greater than that of pure Cu. Again the reason for this lies in the theory of dislocation mobility, rather than the macroscopic elastic stiffness. This will be discussed in Chapter 18.

It is apparent that the elastic properties of metal alloys is a complex subject. Little that is general can be said about it. A factor of some generality is that elastic stiffness is related to the valence electron density. This is not entirely straightforward, however, because it is often not clear what valence number(s) should be assigned to an alloy.

It has been estimated that there are some 5000 identified intermetallic compounds having some 500 different crystal structures. Thus, only a very small sample has been discussed here.

13.7.2 "Hard metals" (metal–metalloid compounds)

In some metals (both simple and transition types), the interstices between the atoms of the structure are large enough to accommodate "metalloid" atoms, including B, C, N, and Si. For example, adding B to Ti forms titanium diboride, TiB_2, and adding C to W forms tungsten carbide, WC. The additional valence electrons of the metalloids increase the valence electron density thereby stiffening the structure. Also, the metalloids form covalent bonds with the metals which tends to reduce dislocation mobility in these compounds. The resulting crystals are very stiff and extremely hard. Therefore they make useful abrasives, wear-resistant surfaces, and they have a variety of other uses.

Hard metal crystals have negligible ductilities at ordinary temperatures because of the low mobilities of dislocations in them. Therefore, they tend to be weak in tension, but exceptionally strong in compression.

A set of six transition metals forms the prototypical carbides, namely, titanium, zirconium, and hafnium of valence four, and vanadium, niobium, and tantalum of valence five. All of these carbides have the same rocksalt crystal structure. Several other elements form carbides (including tungsten, molybdenum, chromium, silicon, and boron), but they do not form a simple set, all of the same structure. Figure 13.9 summarizes the elastic properties of the carbide-forming elements relative to their carbides. The top half of the figure compares the bulk moduli of the metals and their carbides, while the bottom half does the same for the average shear moduli. The bulk moduli of the carbides are roughly three times larger than those of the corresponding metals. The increases of the shear moduli are even larger, nearly six times larger in some cases. The similarities of the patterns of the six bulk and shear moduli are apparent.

Since the shear moduli increase relatively more than the bulk moduli, the ratio G/B is substantially greater for the carbides than for the metals. For the six metals, the average of the G/B ratios is 0.37, while the average value for the carbides is 0.71, almost twice as large. This accounts, at least in part, for the greater rigidity (brittleness) of the carbides compared with the metals.

From the viewpoint of electronic structure, clear trends are not easy to identify. In Figure 13.9, the bulk moduli for the V, Nb, Ta set are all larger than for the Ti, Zr, Hf set; presumably because there is one extra valence electron without much change in the molecular volume for the V, Nb, Ta set, so its valence electron density is larger. However, this simple trend does not persist for the shear moduli.

The nitrides of these six elements that have the same rocksalt crystal structure tend to be less stiff than the carbides. It is not clear whether this results from the larger size

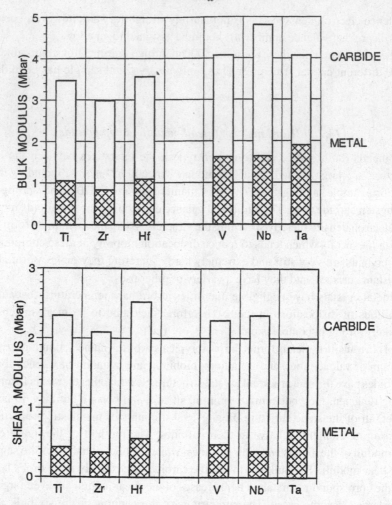

Figure 13.9 Comparing the moduli (bulk at top, shear at bottom) of the six prototype transition metals and their carbides. Note that when carbon is added to these metals, their shear moduli increase by a factor of about three. This increase is believed to be a result of the increased valence electron density, attributing four electrons to carbon.

of the nitrogen atoms, or from occupation of anti-bonding states by the extra valence electron.

The borides (usually diborides) of the six carbide-forming elements have large elastic stiffnesses, but they are difficult to compare because the borides have more than one crystal structure.

In all cases the carbides are substantially more stiff than the corresponding metals. Qualitatively, this is because the valence electron density is increased by formation of the carbide, but it is not clear how many "valence" electrons are contributed by the carbon in the various cases, so there is little more of a simple nature to be said. Theories, based on band structure calculations and empirical potential energies, are discussed elsewhere (Cottrell, 1995; Gubanov, Ivanovsky, and Zhukov, 1994).

13.8 Ionic crystals

13.8.1 Alkali halides

The alkali halides are compounds formed by combining the most electropositive elements with the most electronegative ones. Therefore, they are the salts most likely to contain discrete ions that can be represented by centers of positive and negative charge. Theories of cohesion based on this simple picture formed the early basis of the physics of the solid state, and have been remarkably successful (Chapter 12). This has provided evidence that the point-charge assumption is reasonably valid. Calculations of cohesive energies based on experimental charge distributions found through X-ray scattering agree quite well with the point-charge assumption (Coppens, 1997). The fact that these crystals are soluble in water reinforces the idea that they consist of arrays of point charges. The dielectric constant of water is quite high (≈ 80), so an array of point charges can decrease its energy by dispersing itself in it.

Since the advent of quantum chemistry which brought the realization that electron densities decline exponentially at the outsides of atoms, the repulsive term in the potential energy has been taken to be an exponential function of the distance r_{ij} between ions instead of an empirical inverse power function. Therefore, the expression for the energy of two ions, i and j, is often written:

$$u_{ij}(r) = \mp q^2/r_{ij} + R \exp(-r_{ij}/\rho) \tag{13.16}$$

where R and ρ are constants determined from measured elastic stiffnesses. Using this potential function, Krishnan and Roy (1952) found the following for the shear constants of crystals with the NaCl structure, and lattice parameters, r_0:

$$C_{44} = 0.348 q^2/(r_0)^4 \tag{13.17}$$

This yields the values in Table 13.4 for the most favorable salts. It may be seen that the agreement between the calculated and the measured values is quite good. The expected dependence on the Keyes parameter is also good (Gilman, 1969). However, the theory is not complete as indicated by a small dependence on the cations not included in Equation (13.13). This dependence is partially associated with temperature, and decreases substantially if low temperature values for the constants are used.

In Section 13.4.2 a connection between polarizability and shear stiffness was derived. As a first approximation, stiffness and polarizability are inversely proportional, as indicated by Figure 13.10. The pattern of reciprocal polarizabilities is the same as that of the elastic stiffnesses with LiF most stiff, and CsI least. However, a more detailed examination indicates that other factors are also important. See Figure 13.11 where the individual shear stiffnesses

Table 13.4 *Shear moduli, C_{44} (10 GPa) for some alkali halides*

Crystal	Measured	Calculated
NaCl	1.26	1.30
NaBr	0.97	1.00
KCl	0.63	0.80
KBr	0.51	0.70

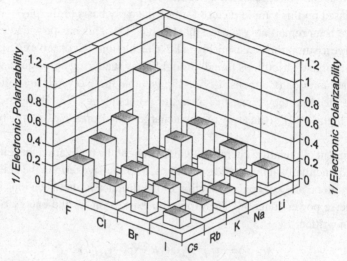

Figure 13.10 Pattern of reciprocal polarizabilities for the alkali halides. LiF is the largest, and CsI the smallest.

are displayed with more precision plotted against the reciprocal polarizabilities. Each group in this figure consists of four halides, F, Cl, Br, and I. It is apparent that there is considerable dependence on both of the specific ions, especially for the C_{44} values.

The spread of Figure 13.11 is reduced somewhat if the data are plotted in terms of the electronegativity difference density (Figure 13.12), but it is not eliminated. The best correlating factor is the same as for the metals, namely, the valence electron density (or, since the alkali halides all have the same valence number, the reciprocal molecular volumes), see Figure 13.13.

Unlike the bulk moduli, the shear moduli of the alkali halides are not reduced to a single universal correlation line by ploting the data versus the electronegativity difference density, see for example the C_{44} data in Figure 13.12.

Other sets of compounds for which there are systematic trends are the oxides and fluorides of the alkaline earth metals. Small systematic sets of data can also be found for some

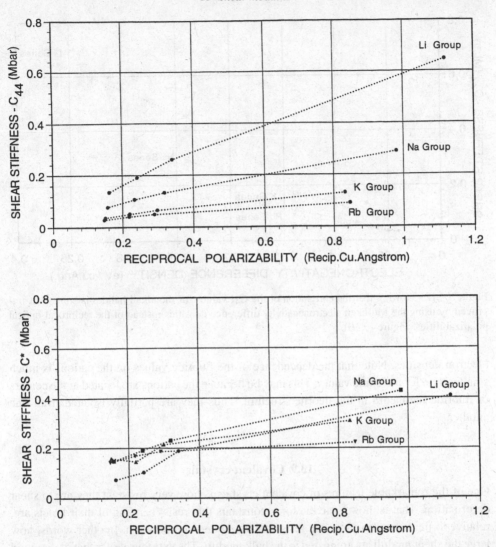

Figure 13.11 Shear stiffnesses, C_{44} and C^*, of the alkali halide groups as functions of their reciprocal polarizabilities (optical). For example, the Li group is LiF, LiCl, LiBr, and LiI.

intermetallic compounds, but the sets are small because the crystal structures are quite variable. Except for the chalcogenides these trends will not be discussed here.

13.8.2 Alkaline earth fluorides and oxides

The shear moduli C_{44} of the fluorides and oxides are displayed as a function of their valence electron densities in Figure 13.14. The oxide values are directly proportional to the

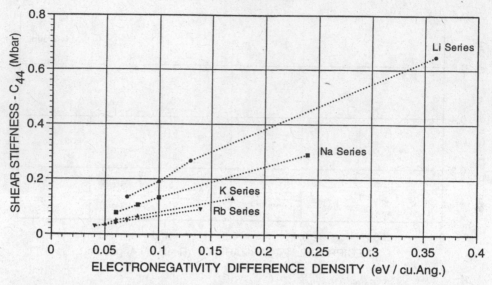

Figure 13.12 Showing that the correlation of the C_{44} values for the alkali halides is somewhat improved by using the Mulliken electronegativity difference densities instead of the reciprocal optical polarizabilities (Figure 13.10).

electron densities. Note that the dependence of the fluoride values on the cations is much smaller than for the oxide values. This may be because the cations are located at the centers of fluoride tetrahedra in the fluorite structure. Thus they are partially isolated from one another.

13.9 Covalent crystals

One of the remarkable features of covalent crystals is how very resistant they are to shear deformation. That is, how large the force constants that resist bending of their bonds are, relative to the force constants that resist stretching and compression. In other words, how large the shear moduli are compared to the bulk moduli. The extreme case is that of diamond where the elastic constants are:

$$C_{11} = 10.8 \text{ Mbar}$$
$$C_{12} = 1.25 \text{ Mbar} \qquad (13.18)$$
$$C_{44} = 5.77 \text{ Mbar}$$

These constants yield $B = 4.42$ Mbar which is smaller than C_{44}, and is also smaller than $C^* = (C_{11} - C_{12})/2 = 4.76$ Mbar. The value of B for diamond is larger than the B for any other element with the possible exception of Os (the minimum reported value for this element is less than 44.2 Mbar, but the maximum is greater). The unique shear stiffnesses give diamond its unusual hardness, thermal conductivity, and optical refractivity. Note that

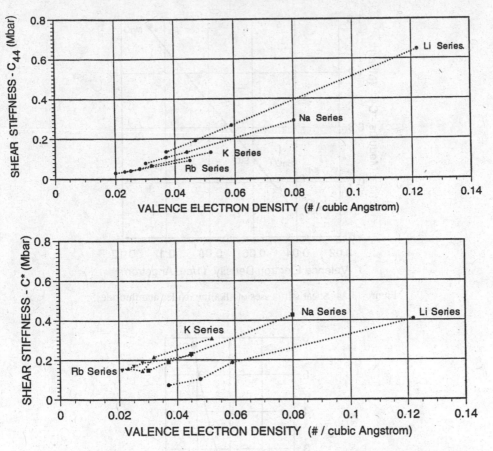

Figure 13.13 Dependences of the shear stiffnesses of the alkali halides on their valence electron densities (reciprocal molecular volumes).

the overall elastic anisotropy is relatively small, $C_{44}/C^* = 1.21$, while the Cauchy ratio is large, $C_{44}/C_{12} = 4.62$.

The unusual case of C_{44} being greater than B is also true for cubic-BN, and for cubic-SiC, which also have strong covalent bonding.

The deviations of the internal force constants from simple radial atom–atom forces directed along the bonds (Cauchy ratio of unity) are illustrated in Figure 13.15 for the Group IV crystals. The Cauchy ratio (C_{44}/C_{12}) varies from about 1.0 for Sn to about 4.6 for diamond. Resistance to bond bending reaches a peak with carbon which has only four electrons in its outer shell and therefore relatively little shielding of its positive nuclear charge.

The shear moduli for crystals with the diamond structure can be estimated with a calculation made in the spirit of the chemical theory known as VSEPR theory, or valence shell electron pair repulsion theory (Atkins, 1991). This is a close relative of the bond charge model (Parr and Yang, 1989). To make this calculation, it is assumed that the negative

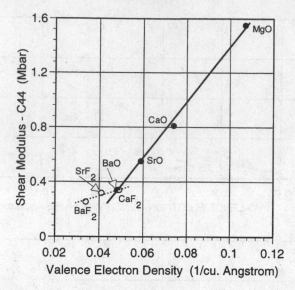

Figure 13.14 Shear stiffnesses of alkaline oxides and fluorides.

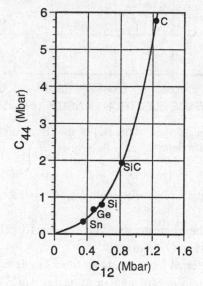

Figure 13.15 Dependence of the Cauchy ratios of the Group IV elements. The ratio is the slope of the line.

charges of each bond act as if they were concentrated at the quarter-points of the bonds. The positive charge is assumed to be concentrated at each nucleus. We focus on a bond (A) that crosses the (111) plane (Figure 13.16). The bond is subject to shear in the xz-plane. This changes the distance between the centers of negative charge. Since each end of the bond of

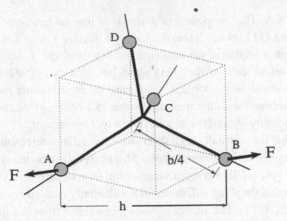

Figure 13.16 Schematic arrangement of partial bond charges near a tetrahedral node in diamond. The node (carbon nucleus) is at the center of the schematic cube. The bonds surrounding the carbon nucleus are A, B, C, and D with the shaded circles representing the partial bond charges. The latter lie at the $b/4$ positions along the bonds of length b. The distance between each pair of bond charges is h, and F is the principal force between each pair of partial bond charges.

interest is anchored by a tripod of other bonds (BCD) which are very stiff in their stretching modes, it is assumed that the compliances of the tripods can be neglected. In diamond, the valence band is just filled with electron spin-pairs. Since they have the same charges, the electrons avoid one another. They can do this quite well if their motion is correlated and they spend more time at the one-quarter and three-quarter points along the bonds than at the mid-points.

In the diamond structure, the (111) planes are the most widely separated ones. Since these planes have three-fold symmetry axes, there is only one shear modulus, C^{111}, associated with them. In terms of the usual elastic constants, Equation (13.18), this shear constant is given by:

$$C^{111} = [3C_{44}(C_{11} - C_{12})]/[4C_{44} + (C_{11} - C_{12})] \qquad (13.19)$$

The numerical value is $C^{111} = 5.07$ Mbar which is 14.4% larger than the bulk modulus, $B = 4.43$ Mbar.

In Figure 13.16, a set of partial tetrahedral bonds surrounding a carbon atom at the center of a schematic cube is shown. Focus on the bond pair, A and B. Each of the bonds has concentrations of charge (the shaded circles) at the 1/4, 3/4 positions along its length, b. The two concentrations nearest the nodal carbon atom will have the largest effect on the stiffness of the angle between A and B. The interactions of A with C and D will be the same, but in different directions, so they will be neglected. Although the magnitudes of the charge concentrations (the shaded circles) are not accurately known, they must be large in order to be seen clearly by means of X-ray scattering. Therefore, let them be unit electron charges. Then the Coulombic repulsive force, F, will be $F = q^2/h^2$ where h is the distance between the centers of charge concentration. Since the bonds are 1.54 Å long and the bond angle is

109.47°, $h = 0.628$ Å. The component of F that is of interest lies perpendicular to bond A which is normal to a (111) plane. Then the resolved force is $f = F \cos 54.8° = 0.577F$.

In order to obtain a modulus, the area (over which the force f acts) is needed. In this structure, a unit surface cell on the (111) plane has an area of 5.51 Å^2. Therefore, the estimated shear modulus is 6.1 Mbar. Considering the approximate nature of the model, this is in good agreement with the measured value, 5.1 Mbar, given above. Reducing the charge concentrations by about 10% would yield exact agreement.

It is concluded that the unusually high shear stiffness of diamond results from the bimodal distribution of charge along its covalent bonds. This charge distribution is a result of electron correlation which is especially strong in the diamond form of carbon. As would be expected from the simple model, the shear stiffnesses of tetrahedrally bonded crystals are sensitive to the ionicity (Phillips, 1973). The greater is the charge transfer from one ion to another in AB compounds, or the less localization at bond centers in homopolar crystals, the smaller are the shear stiffnesses of the bonds that cross the (111) plane. The shear constant that describes this stiffness, C^{111}, can be normalized by dividing it by the bulk modulus and plotting the reduced values, shown in Figure 13.17. With the exception of the case of diamond, this figure shows that there is a good correlation between relative shear stiffness and ionicity. However, the fact that diamond is exceptional in this figure suggests that ionicity may not be the best parameter for correlating the various crystals.

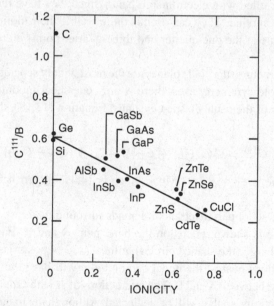

Figure 13.17 Effect of ionicity on shear stiffness. The ionicity values are those of Phillips (1973). The stiffness values, C^{111}, are for shearing on the octahedral planes (111) of the diamond structure. These have been normalized by dividing them by the values of the bulk moduli, B. Note that the values for diamond do not fall on the trend line.

A parameter that may correlate covalent shear stiffnesses better than ionicity is polariz-ability. However, a still better parameter in the case of solids might be one suggested by the work of Brinck, Murray, and Politzer (1993) who showed that although the polarizabilities of a variety of molecules are approximately proportional to their molecular volumes, V, they correlate even more closely with (V/I_{ave}) where I_{ave} is the average ionization energy over the surface of a molecule. An approximate analog of I_{ave} is the minimum band gap (the internal ionization energy). Only the shear stiffnesses of the bonds across the octahe-dral (111) planes are considered. These are obtained from the standard elastic constants by Equation (13.18). Data for the Group IV elements are plotted against $I_{ave}/V_m = E_g/V_m$ in Figure 13.18. Data for the isoelectronic III–V compounds also display a good linear correlation. Plotting C_{44} this way also yields a linear variation.

· The mechanism connecting the gap density (units of stress) and the shear stiffness (also units of stress) may be that in order to change the shapes of these crystals their bonding electrons must cross the gaps into excited states that have non-spherical amplitude func-tions. This internal ionization process reduces the localization of the valence-electron pairs, thereby tending to increase the polarizability, and decrease the shear stiffness as Figure 13.17 indicates.

· The most simple correlating parameter for the bulk moduli of the covalent crystals is the valence electron density, and this is also true for shear stiffness (Figure 13.19).

Ionicity affects the stabilities of covalent crystals. Highly ionic crystals prefer the dense rocksalt structure (coordination number 6) to the open diamond, zinc blende, or wurtzite structures (coordination number 4). Relative shear stiffnesses which are affected by ionicity

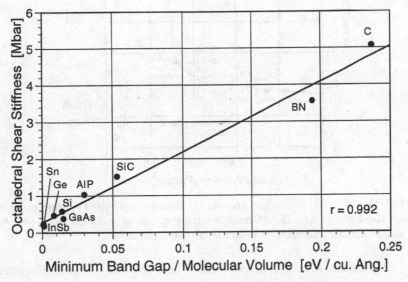

Figure 13.18 Correlation between the octahedral shear stiffness and the band gap density (gap/unit volume) for Group IV elements, and their isoelectronic III–V compounds.

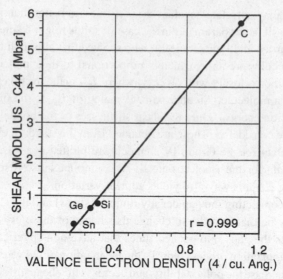

Figure 13.19 Showing the linear dependence of C_{44} on the valence electron density.

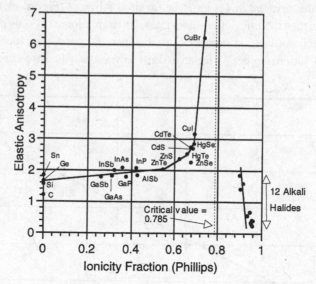

Figure 13.20 Map of elastic anisotropies and ionicity fractions for Group IV elements, III–V compounds, and II–VI compounds. The structural instability at ionicity fraction 0.785 is shown, where the coordination number changes from 4 (tetrahedral) to 6 (octahedral).

(Figure 13.17) are also important as a measure of the stabilities of crystal structures. This was pointed out first by Zener (1948) for the body-centered cubic structure, particularly for the case of the alkali metals where $C^{\star}$ is small compared with C_{44}. Zener's conjecture was confirmed experimentally by Barrett (1947) for Li and Na. It is also true for the zinc

Table 13.5 *Effect of isotopic composition on the elastic stiffness of diamond (values in Mbar)*

Isotope	C_{11}	C_{12}	C_{44}
^{12}C (100%)	10.799	1.248	5.783
^{13}C (99.2%)	10.792	1.248	5.776
Difference (%)	0.065	0.0	0.121

Data from Ramdas *et al.*, 1993.

blende ↔ rocksalt transformation in semiconductor compounds (Phillips, 1973). Figure 13.20 is a plot of the anisotropy parameter (ratio of the two shear stiffnesses of cubic crystals) versus the ionicities of various binary compounds (Nikanorov and Kardashev, 1985). Ionicity is dependent on the electronegativity difference between the components of a crystal. It is a measure of the ionic character (i.e., the tendency to form the rocksalt structure, and to dissolve in water).

Measuring down from the vacuum energy level, the electronegativity (Mulliken scale) is the average of (1) the ionization energy of an atom (the difference between the vacuum and the HOMO levels); and (2) the electron affinity (the difference between the vacuum and the LUMO levels). Thus it lies halfway between the LUMO and HOMO levels.

The critical value of the ionicity where the preferred structure changes from the zinc blende structure (coordination number 4) to the rocksalt structure (coordination number 6) is 0.785 (Phillips, 1973). It coincides with the spike in the anisotropy parameter which occurs when C^* becomes small compared to C_{44}.

13.10 Isotope effect

A subtle effect of isotopic composition has been demonstrated for diamond (Ramdas *et al.*, 1993). Natural diamond consists mostly of the ^{12}C isotope, but synthetic diamonds with various isotopic compositions have been grown, including nearly pure ^{13}C. This causes small changes in the physical properties including the elastic stiffnesses. Table 13.5 lists the very small observed changes. The change in the shear modulus C_{44} is about twice as large as the change in C_{11}.

13.11 Quasicrystals

These are metallic alloys which contain two, or more, kinds of small cells of atoms that are packed together in a non-periodic way, yielding "crystals" with five-fold or ten-fold symmetry. It does not appear that the quasi-periodic structures of these materials introduce any qualitatively new behavior. According to the measurements of Tanaka, Mitarai, and

Koiwa (1996), the behavior is similar to that of periodic crystals of similar compositions. This simply confirms that it is the local atomic interactions that determine the macroscopic elastic behavior, not long-range periodicity.

13.12 Polymers

Commodity polymers such as polyethylene, nylon, and polymethacrylate have relatively small elastic stiffnesses. These are largely determined by the weak London dispersion forces between the molecules. Therefore, these materials are not used where high stiffness is needed. Functional polymers, however, are sometimes exceptionally stiff. These are made either by drawing fibers until their molecules are fully aligned along their fiber axes, or by "blowing" films close to their fracture limits to obtain biaxial alignment, or by reactive spinning of polyaromatics to obtain highly aligned, and very stiff, fibers. Highly oriented fibers are, of course, very anisotropic.

Often, as in polyethylene, the molecules have a puckered or "zig-zag" configuration. Or, in biopolymers, including proteins and nucleic acids, the polymers are coiled. Then when applied forces stretch these molecules, the resulting extensions come from a combination of straightening of the bond angles, lengthening of the bond lengths, and bond torsions. For small strains, each of these processes is elastic, and the stresses and strains can be calculated by means of molecular mechanics, using spring constants derived from spectroscopic data. In laboratory specimens, there are additional forces between the aligned molecules, but these are relatively weak and are neglected without introducing large errors. The calculations are discussed at some length in the books by Kelly and Macmillan (1986), and Ward (1997). Table 13.6 shows good agreement between the calculated and measured values. Graphite is included as a benchmark.

The transverse moduli (C_{11} and C_{22}) are much smaller than the axial (C_{33}) moduli. For example, the average transverse modulus for polyethylene is about 0.13 Mbar which is 25 times smaller than its chain modulus, C_{33}.

Table 13.6 *Young's moduli of polymer fibers*

Polymer	Condition	Modulus C_{33} (Mbar)	
		Measured	Theory
Polyethylene	zig-zag chain	3.4	3.4
Polyvinylalcohol	zig-zag chain	2.4	2.5
Polypropylene	helical chain	0.4	0.4
Polyparaphenylene terephthalamide (Kevlar)		1.8	1.8
Nylon-6		1.7	2.0
Graphite	scroll axis	10.2	—

Data from Ward, 1997.

13.12.1 Intermolecular stiffness (London forces)

The atoms in all solids are in various states of vibration. Even if the temperature is absolute zero, each mode of oscillation has half of a quantum of vibrational energy. The vibrational oscillators (dipoles) emit and absorb photons and phonons constantly. Thus particles are constantly being exchanged between the oscillators, and this generates weak attractions between them. These are the forces, for example, that bind polymer molecules together (intermolecular forces) to make bundles or crystals. They are much weaker than the covalent forces (intramolecular forces) that bind monomers into chains and form cross-links. The forces between non-polar molecules are sometimes considered to occur through the dipole moments that may be generated by mutual self-induction.

If molecules with permanent dipole moments are present (polar molecules) they can interact electrostatically with other polar molecules, or with charged atoms and other localized charges.

Since the dipolar forces are always attractive, those of the individual molecules in macroscopic bodies add up to form small, but finite, forces between any two solid bodies (in addition to gravitational forces). These have been directly measured, and conform to theoretical expectations (Derjaguin, Churaev, and Muller, 1987).

The forces are not generated instantaneously because the exchange photons have finite speeds, the maximum being the speed of light (3×10^{10} cm/s). Therefore, for solid bodies widely separated, the forces are somewhat reduced by the retardation in the response of a receiving dipole relative to the time at which a photon leaves a sending dipole. This is known as the Casimir effect. The distance at which retardation becomes important is $\sim c/\nu$ where c is the speed of light and ν is the oscillator frequency. At distances d greater than this, the dipole–dipole forces decay in proportion to $1/d^7$. At shorter distances, the forces are somewhat larger and decay in proportion to $1/d^6$. In this domain, the forces are known as London forces (Langbein, 1974).

As would be expected, the dipole–dipole forces depend not only on the distance, d, but also on the orientations and other properties of the oscillators. In particular they depend on the ease with which the oscillators become polarized, that is, on their polarizabilities. A simplified expression for the polarizability, α, of a quantized oscillator is (Atkins, 1983):

$$\alpha = \frac{2}{m}\left(\frac{qh}{2\pi\Delta}\right)^2 \qquad (13.20)$$

where q is the electron charge, h is Planck's constant, m is the electron mass, and Δ is the energy difference between the ground s-state and first excited p-state. Thus the polarizability increases rapidly with decreasing excitation energy, Δ, and so does the dipole–dipole attraction.

13.13 Atomic vibrations

Ledbetter (1991) has pointed out that estimates of the characteristic Debye temperatures of 24 cubic elements can be markedly improved by using shear moduli instead of the conventional practice of using bulk moduli. From the relation $k\Theta = h\nu$, the Debye temperature is related to the characteristic frequency, ν, using the Boltzmann constant, k, and the Planck constant, h. The mean frequency, ν_m, is given by: $\nu_m = s_m / V_a^{1/3}$, where s_m is the mean speed of elastic waves, and V_a is the atomic volume.

There are two wave speeds on average in polycrystalline cubic material: s_l for longitudinal waves, and s_t for transverse waves. The mean velocity, s_m, which determines Θ, is given by: $s_m^{-3} = (1/3)(s_l^{-3} + 2s_t^{-3})$. The two velocities obey the relations: $s_l^2 = (B + 4G/3)/\rho$ and $s_t^2 = G/\rho$, where B is the bulk modulus, G is the shear modulus, and ρ is the mass density. These relations, plus a typical Poisson ratio of 1/3, lead to:

$$(d\Theta/\Theta) = (1/51)[(dB/B) + (49/2)(dG/G)]$$

so G has a much larger effect on Θ than does B. Furthermore, compared with more exact calculations, the use of G instead of B improves the variance of the calculated Θ from an average of 24% to an average of 2%. Needless to say, this is important for the theory of specific heat.

References

Anderson, O.L. and Isaak, D.G. (1995). Elastic constants of mantle materials at high temperatures. *Mineral Physics and Crystallography – A Handbook of Physical Constants*, AGU Reference Shelf 2, p. 64. Washington, DC: American Geophysical Union.

Atkins, P.W. (1983). *Molecular Quantum Mechanics*, 2nd edn. Oxford: Oxford University Press.

Atkins, P.W. (1991). *Quanta*, 2nd edn., p. 377. Oxford: Oxford University Press.

Barrett, C.S. (1947). *Phys. Rev.*, **72**, 245.

Blatt, J.M. and Weisskopf, V.F. (1991). *Theoretical Nuclear Physics*, p. 23ff. New York: Dover.

Bonin, K.D. and Kresin, V.V. (1997). *Electric-Dipole Polarizabilities of Atoms, Molecules and Clusters*. Singapore: World Scientific.

Brinck T., Murray, J.S., and Politzer, P. (1993). Polarizability and volume, *J. Chem. Phys.*, **98** (5), 4305.

Coppens, P. (1997). *X-Ray Charge Densities and Chemical Binding*. New York: Oxford University Press.

Cottrell, A.H. (1988). *Introduction to the Modern Theory of Metals*. London: The Institute of Metals.

Cottrell, A.H. (1995). *Chemical Bonding in Transition Metal Carbides*. London: The Institute of Materials.

Cottrell, A.H. (1997). The art of simplification in materials science, *Mater. Res. Soc. Bull.*, **22** (5), 15; correction, **22** (9), 10.

Derjaguin, B.V., Churaev, N.V., and Muller, V.M. (1987). *Surface Forces*, trans. V.I. Kisin and J.A. Kirchener. New York: Consultants Bureau.

Feynman, R.P., Leighton, R.B., and Sands, M. (1964). *Feynman Lectures on Physics*, Volume II. Reading, MA: Addison-Wesley.

Frank, F.C. (1992). Body-centered cubic – a close-packed structure, *Philos. Mag. Lett.*, **65** (2), 81–84.

Fuchs, K. (1935). *Proc. R. Soc. London, Ser. A*, **151**, 585.

Gilman, J.J. (1966). Monocrystals in mechanical technology, Campbell Memorial Lecture, *Am. Soc. Met. Trans. Q.*, **59** (4), 597.

Gilman, J.J. (1969). *Micromechanics of Flow in Solids*, p. 33. New York: McGraw-Hill.

Gilman, J.J., Cunningham, B.J., and Holt, A.C. (1990). Method for monitoring the mechanical state of a material, *Mater. Sci. Eng.*, **A125**, 39.

Gilman, J.J. (1997). Chemical and physical hardness, *Mater. Res. Innovat.*, **1**, 71.

Gubanov, V.A., Ivanovsky, A.L., and Zhukov, V.P. (1994). *Electronic Structure of Refractory Carbides and Nitrides.* Cambridge: Cambridge University Press.

Harrison, W.A. (1980). *Electronic Structure and the Properties of Solids.* San Francisco, CA: W.H. Freeman.

Hecker, S.S., Rohr, D.L., and Stein, D.F. (1978). Brittle fracture in iridium, *Metall. Trans. A*, **9**, 481.

Kelly, A. and Macmillan, N.H. (1986). *Strong Solids*, 3rd edn. Oxford: Oxford University Press.

Keyes, R.W. (1962). Elastic properties of diamond-type semiconductors, *J. Appl. Phys.*, **33**, 3371.

Krishnan, K.S. and Roy, S.K. (1952). *Proc. R. Soc. London, Ser. A*, **210**, 481.

Langbein, D. (1974). *Theory of van der Waals Attraction*, Springer Tracts in Modern Physics, Volume 72. Berlin: Springer-Verlag.

Ledbetter, H. (1991). Atomic frequency and elastic constants, *Z. Metallkd.*, **82** (11), 820.

Love, A.E.H. (1944). *A Treatise on the Mathematical Theory of Elasticity*, 4th revised edn., p. 8. New York: Dover Publications.

Machlin, E.S. and Whang, S.H. (1978). Atomic shear constant, Born repulsion charge cloud, and nuclear quadrupole coupling constant in the A!5 structure, *Phys. Rev. Lett.*, **41** (20), 1421.

Mahan, G.D. and Subbaswamy, K.R. (1990). *Local Density Theory of Polarizability*, Chapter 4, p. 87. New York: Plenum Press.

Nikanorov, C.P. and Kardashev, B.K. (1985). *Elasticity and Dislocational Inelasticity of Crystals* (Uprugost i dislokationalya neuprugost kristallov), p. 98. Moscow: Nauka.

Parr, R.G. and Yang, W. (1989). *Density Functional Theory of Atoms and Molecules*, p. 229ff. Oxford: Oxford University Press.

Pauling, L. (1960). *The Nature of the Chemical Bond*, 3rd edn. Oxford: Oxford University Press.

Pettifor, D.G. (1995). *Bonding and Structure of Molecules and Solids.* Oxford: Clarendon Press.

Phillips, J.C. (1973). *Bonds and Bands in Semiconductors*, p. 62. New York: Academic Press.

Ramdas, A.K., Rodriguez, S., Grimsditch, M., Anthony, T.R., *et al.* (1993). Effect of isotopic constitution of diamond on its elastic constants, *Phys. Rev. Lett.*, **71** (1), 189.

Simmons, G. and Wang, H. (1971). *Single Crystal Elastic Constants: A Handbook.* Cambridge, MA: MIT Press.

Tanaka, K., Mitarai, Y., and Koiwa, M. (1996). Elastic constants of Al-based icosahedral quasicrystals, *Philos. Mag. A*, **73** (6), 1715.

Timoshenko, S.P. (1983). *History of the Strength of Materials.* New York: McGraw-Hill, 1953. Republished by Dover Publications, New York.

Ward, I. (Editor) (1997). *Structure and Properties of Oriented Polymers.* London: Chapman & Hall.

Zener, C. (1948). *Elasticity and Anelasticity of Metals.* Chicago, IL: University of Chicago Press.

14

Entropic elasticity (polymers)

14.1 Introduction

Since both crystals and polymers are constructed of repetitive units (atoms or molecules), it is difficult to differentiate them. Roughly, crystals may be taken to be arrays of repetitive units in which repetition occurs in three dimensions. Polymers may be taken to be arrays that are repetitive in one dimension. However, there are two-dimensional as well as one-dimensional crystals. And there are two-dimensional, and in some cases, three-dimensional polymers. Thus, the distinction is not clean-cut.

Another classification approach is to say that crystals tend to be inorganic in character, whereas polymers tend to be organic. In a majority of cases this is true, but by no means all cases.

Covalent bonding within one-dimensional chains of atoms (or molecules) has already been discussed in Chapter 9. Most real polymers are two-dimensional, however, at least to the extent that the bonds form "zig-zag" patterns rather than straight lines. Thus, it is not possible to discuss the extension or contraction of polymer chains in terms of radially directed forces alone. Distortion of planar angles must be considered, and in most cases changes of rotational angles. This latter degree of freedom allows the contracted configuration to be that of a helical coil. Purely planar molecules are rare.

Intramolecular bonds are usually covalent, whereas intermolecular ones are usually due to London forces (in most cases Casimir forces can be neglected), but they can also be due to hydrogen exchange, or if the polymers contain polar groups they can be due to the interactions of pairs of permanent dipoles, or permanent/induced dipole pairs.

Another new feature that appears for polymers because of their high compliances (this lies between the compliance of a gas and that of a hard crystal like iron) is that they exhibit a large thermo-elastic effect (rubbery elasticity). This is often called *entropic elasticity*, as contrasted with the more usual *enthalpic elasticity*. However, it can equivalently be viewed as a thermokinetic effect (Weiner, 1987). An important aspect of this is that for low frequencies (or relatively high temperatures) of deformation, polymers have a "relaxed" elastic modulus that is small, while at high frequencies (or relatively low temperatures) of deformation they have an "unrelaxed" elastic modulus that is much larger. The latter case will be discussed here first.

174

14.2 Enthalpic stiffness

In structural materials, the forces that generate elastic stiffness are predominantly electrody-namic, that is, a combination of electrostatics and dynamic exchange forces associated with various quasi-particles, including phonons, photons, electrons, and protons. These forces determine the bending and stretching spring constants of the covalent bonds *between atoms*. They also determine the van der Waals binding *between molecules* that are extended in one (polyethylene), or two (graphite sheets), dimensions. The electrodynamic forces exist at all temperatures because all quantized oscillators have zero-point energies, including the virtual oscillators of the vacuum electromagnetic fields.

The origins of electrodynamic cohesive forces have been outlined in Chapters 9–11. Since they all derive from distributions of electronic charge, electrostatics can be used to calculate them if the distribution is known. From Coulomb's Law, electrostatic forces f are proportional to the square of the ratio of the electronic charge q to the distance between two given charges r, that is: $f \sim q^2/r^2$. But elastic moduli, M, are determined by the curvature of the force–distance curve, so they have the dimensions of stresses, $\sigma = f/r^2$ or:

$$M \sim q^2/r^4 \tag{14.1}$$

This is known as the Keyes parameter, $K = q^2/r^4$. Keyes (1962) showed that this factor rationalizes the data for a large number of semiconductors.

14.3 Entropic stiffness

Molecules called elastomers (rubber-like) are polymeric, consisting of segments connected together by joints that rotate freely. Although the strong bonds within the segments deform very little under applied loads, the chains of segments can be either extended, or contracted, depending on the temperature. At high temperatures, thermal motions buffet the segments, tending to randomize their positions and thereby contracting the chains. At low temperatures, the segments relax so small forces can cause the chains to become extended. It is apparent from this mechanism that the resistance to extension (the stiffness) increases with increasing temperature. This is the opposite of enthalpic elasticity where the stiffness slowly decreases with increasing temperature.

The two forms of elasticity are related to the two terms of the expression for the Gibbs free energy, $F = U - TS$, where U is the internal energy, or enthalpy, of the material, T is the temperature, and S is the entropy of the material. A change in this energy per unit applied strain ϵ (per unit volume) produces a stress whose magnitude is determined by an elastic modulus M. Thus:

$$\sigma = M(\partial F/\partial \epsilon) = M[(\partial U/\partial \epsilon) - T(\partial S/\partial \epsilon)] \tag{14.2}$$

At zero temperature, the first term of the right-hand side describes "enthalpic" elasticity where the resistance to deformation comes from distortion (stretching and bending) of the positions of the atoms relative to one another. In elastomers, however, the nearly rigid segments of the polymer chains can readily change the angles between adjacent segments without significantly changing the positions of the atoms within the segments. Fluctuations of thermal energy cause the elastomer segments to take up random orientations, that is, to increase their entropy. This creates a virtual stress according to the last term of Equation (14.2). This virtual stress resists tensile deformation which tends to align (or anti-align) the segments with the axis of the principal stress.

14.4 Rubbery elasticity

The fact that rubber elasticity is related to finite changes in the molecular structure of the material rather than the infinitesimal changes that occur in the elastic deformation of metals was dramatically shown by Hock (1925). He found that if rubber is strongly extended, then cooled in liquid nitrogen and shattered by a sharp blow, the fractured pieces are fibrils. In contrast, if the rubber has not been extended, just cooled and struck, it breaks into conchoidal pieces. Clearly, when rubber is unextended its microscopic structure is randomized, and the molecules become oriented parallel to the extension axis when it is axially deformed. Since the segments of the polymeric molecules can rotate nearly freely at their junctions, the molecules can lengthen and become oriented by relatively small applied forces, so their bond lengths and bond angles do not need to change much during the orientation process.

As the temperature of a piece of rubber is raised, the vibrational amplitudes of its molecular segments increase. For small extensions of the molecules this exerts only small forces on their ends. But, as they become straightened, the end forces increase. Therefore, the time average of the end forces (that is, the macroscopic stress) increases with the amount of extension and the temperature. Thus, the behavior is similar to that expected for an "anti-gas" of segments, and it can be described quantitatively by theory that is similar to but an inverse of the kinetic theory of gases (Gurney, 1966). For an ideal gas, changes in volume at constant temperature do not cause the internal energy to change, but increasing the volume increases the entropy, thereby decreasing the temperature. For rubbery molecules, stretching the length (at a constant number of segments) decreases the entropy, thereby increasing the temperature. This can be felt directly by quickly stretching a rubber-band, and then touching it to one's upper lip.

The motions of the segments of the rubber molecules are not entirely ideal, however, because rotations of the bonds do require small amounts of energy which is converted into other modes of vibration (heat). The gas-like behavior is also limited, of course, by the fact that the molecules become fully extended for large macroscopic extensions. Then further deformation requires changes of the bond angles and lengths, and the stiffness increases considerably.

In a gas, collisions of the particles with the container walls create pressure. In the case of the rubber molecules, collisions of the segments with each other pull on the ends of the molecule drawing them together on average.

Let us focus our attention on a rubbery polymer molecule that consists of N segments each of length λ. Then, if the molecule is fully extended, its length is $L = \alpha N \lambda$ where α is a geometric factor (of order unity) that depends on the structure of the molecule. For simplicity, set $\alpha = 1$, so $L = N\lambda$ which is the length of the molecule when it is fully extended. It is then said to be fully ordered, and its order parameter, ω, is unity. When it is completely randomized, $\omega = 0$ and the distance, δ, between its ends is the same as the distance from the origin of a random walk of N steps, each of length λ: $\delta = (L\lambda)^{1/2}$. For $L \gg \lambda$, this is just $\delta = L^{1/2}$.

For a molecule with its ends held a distance δ apart, its total energy, $H(q, p, \delta)$, is a function of the coordinates, q, of its segments, the momenta, p, of its segments, and δ. At equilibrium, with the temperature being T, its partition function is then:

$$Q(\delta, T) = \int \int \{\exp[-H(q, p, \delta)/kT]\}\, dq\, dp \tag{14.3}$$

and statistical mechanics (Gurney, 1966) yields for the force, f, on the ends:

$$f = -kT[\partial(\log Q)/\partial\delta] \tag{14.4}$$

or:

$$f = [\partial H/\partial\delta]_{q,p} \tag{14.5}$$

Thus the force increases with increasing temperature in accordance with measurements. The proportionality is given by k, the Boltzmann constant.

This method of deriving an expression for the force has been shown by Weiner (1987) to be equivalent to the more usual way of deriving it in terms of the entropy. However, this method has the advantage of running more parallel to the kinetic theory of gases than the conventional method, and it is more direct.

The force modulus m (ratio of force to extension, dyn/cm) increases with temperature, T, and decreases with polymer length, L, and segment length, λ. The proportionality constant is $3k$:

$$m = 3kT/L\lambda \tag{14.6}$$

By means of the atomic force microscope these moduli have been measured directly for a few molecules.

The stress modulus M (Young's modulus) of an aggregation of molecules is of more interest than the force modulus for the practical applications of polymers. It is given by an expression similar to Equation (14.6):

$$M = 3kT/\lambda^3 = 3kTn \tag{14.7}$$

where n is the concentration of molecular segments (Grossberg and Khokhlov, 1997). At room temperature (300 K), kT is 4.1×10^{-14} erg. The elastic modulus of rubbery polystyrene is about 10^6 dyn/cm^2. Therefore, according to Equation (14.7), the expected segment length in polystyrene is 50 Å which is about right.

It is not expected that Equation (14.7) will be exactly right because there will be London forces resisting deformation in addition to the entropic forces.

References

Grossberg, A.Yu. and Khokhlov, A.R. (1997). *Giant Molecules*. New York: Academic Press.

Gurney, R.W. (1966). *Introduction to Statistical Mechanics*, p. 62. New York: McGraw-Hill, 1949. Republished by Dover Publications, New York.

Hock, L. (1925). *Z. Electrochem.*, **31**, 404.

Keyes, R.W. (1962). Elastic properties of diamond-type semiconductors, *J. Appl. Phys.*, **33**, 3371.

Weiner, J.H. (1987). Entropic versus kinetic viewpoints in rubber elasticity, *Am. J. Phys.*, **55**, 746; see also *Statistical Mechanics of Elasticity*, Wiley-Interscience, New York, 1983.

15

Universality and unification

A common theme links chemical, mechanical, and optical hardnesses. It is the gap in the bonding energy spectrum of materials. This gap determines molecular stability, and therefore "chemical hardness" (Pearson, 1997). Chemists call it the LUMO–HOMO gap between anti-bonding and bonding molecular orbitals (LUMO lowest unoccupied molecular orbital, HOMO highest occupied molecular orbital). Physicists call it the band gap between the conduction and valence energy bands (Burdett, 1995). It forms a basis for various properties, including chemical reactivity, elastic stiffnesses, plastic flow resistance (dislocation mobility), crystal structure stability, and optical polarizability (refractive indices). Some of the ways in which this unifying parameter connects chemical, mechanical, and optical properties will be discussed in this chapter.

Another unifying factor is the fact that, for axial bonds between atoms, a simple universal binding energy relationship (called the UBER) exists (Rose, Smith and Ferrante, 1983). Many bonding problems can be reduced to this single relationship which has the form of the "Rydberg function", $f(x) = Ax[\exp(-Bx)]$, and is closely related to the Morse potential. Unfortunately, there is no corresponding universal description for shear deformations.

Partly through coincidence, the name "hardness" which materials engineers have used for centuries to describe mechanical stability was chosen by chemists in recent decades to describe atomic and molecular stability (Pearson, 1997). Happily, the same parameter applies for both chemical and physical properties. It also determines the optical "hardness" of transparent solids.

In solid mechanics, hardness means the resistance to deformation, both elastic and plastic. The particular properties are the bulk modulus which measures the resistance to elastic volume changes, the shear modulus which measures the resistance to elastic shape changes, and dislocation mobilities which determine rates of plastic shape changes. Thus, annealed brass in which dislocations are mobile is said to be "soft", but in the same brass, the dislocations can be immobilized by cold-rolling until it is "hard" enough to be used for springs. An additional property is the critical stress needed to induce a structural transformation, for example, the transformation of silicon from the diamond to the β-tin crystal structure under uniaxial pressure. Still another example is the ease with which twin formation occurs.

Optical hardness is associated with the electronic polarizability which determines the refractive index, and is largely determined itself by the LUMO–HOMO gap (Harrison, 1989).

The idea of chemical hardness is of recent origin. It started in the hands of Pearson (1963) as a means for classifying acids, bases, and their reactions. He proposed that "hard" acids are those with low polarizabilities (stable electron distributions), while "soft" acids are the opposite, similarly for bases. Then he pointed out that "hard" acids tend to react with "hard" bases, and "soft" with "soft". Furthermore, the hard–hard reaction products tend to be the most stable, leading to the "principle of maximum hardness" (Pearson, 1993). These ideas are related to the result of quantum mechanics that strong molecular orbitals form only between valence-electron energy levels that have approximately equal energies.

Chemical hardness, η, was given a quantitative basis by Parr and his associates who used the density functional version of quantum chemistry to connect it to ionization energies, I, and electron affinities, A (Parr and Yang, 1989). It is defined as:

$$\eta = (I - A)/2$$

Thus, it is related to the Hubbard energy gap, $(I - A)$, which plays an important role in the theory of the Herzfeld–Mott metal/insulator transition (Edwards and Rao, 1995). It is also related to the Mulliken electronegativity, $\chi = (I + A)/2$, and is equal to half the LUMO–HOMO energy gap (half the band gap in solids). Finally, it is also related to electronic polarizabilities (Brinck, Murray, and Politzer, 1993).

Five aspects of mechanical behavior are related to chemical hardness. They are: (1) the bulk moduli; (2) the shear moduli; (3) plastic resistance (dislocation mobility); (4) compression-induced phase transformations; and (5) shear-induced chemical reactions.

15.1 Bulk modulus

A thermodynamic formalism due to Pearson (1993) and Yang, Parr, and Uytterhoeven (1987) extends the definition of the *electronic* chemical potential by inference with the similar thermodynamic quantity. The chemical hardness, η, is the derivative of the electronic chemical potential, μ, with respect to the number of electrons, N, in a system, that is, $\eta = (\partial\mu/\partial N)/2$. But $\mu = \partial E/\partial N$, so:

$$\eta = \frac{1}{2}\left(\frac{\partial\mu}{\partial N}\right)_c = \frac{1}{2}\left(\frac{\partial^2 E}{\partial N^2}\right)_c \tag{15.1}$$

where the subscript, c, means that the coordinates of the nuclei are held fixed. By analogy, the corresponding thermodynamic relation is:

$$\mu = \left(\frac{\partial\mu}{\partial N}\right)_{T,V} \tag{15.2}$$

where μ is the thermodynamic chemical potential, N is the number of atoms in volume, V, and T is fixed at zero. From the Gibbs–Duhem relation, and the definition of the bulk stiffness ($B = -V(\partial P/\partial V)$ where P is the pressure), this yields:

$$\left(\frac{\partial \mu}{\partial N}\right)_{V,T} = \frac{-V^2}{N^2}\left(\frac{\partial P}{\partial V}\right) = \frac{BV}{N^2} \tag{15.3}$$

and taking $N = 1$, and dropping the arbitrary 2 in the definition of chemical hardness, we find that the bulk stiffness is proportional to the chemical hardness density η^*. That is:

$$B = \beta(\eta/v_a) = \beta\eta^* \tag{15.4}$$

where v_a is the atomic volume, and β is a constant of order unity.

The bulk moduli for the simple metals are related to the chemical hardnesses of these metals as was shown in Figure 12.11. This is, of course, just an aspect of the connection between valence electron densities and bulk moduli; or put another way, it is an aspect of the Schrödinger pressure of the electrons. The data for the volumes came from Kittel (1976), the chemical hardnesses from Parr and Yang (1989), and the bulk moduli from Rasky and Milstein (1986). Equation (15.4) is verified, with different values of β for each of the three groups, consistent with their different valences.

15.2 Shear stiffness

Since chemical hardness and polarizability are connected (the smaller the polarizability, the greater the chemical hardness), and, as was shown in Chapter 13, shear stiffness can be expressed in terms of inverse polarizability, it is not surprising that the shear moduli and the chemical hardness are related (Gilman, 1997).

This is consistent with other knowledge of the electronic basis of elastic stiffness. For example, it is well known that the elastic stiffnesses of covalently bonded solids can be normalized by means of the "Keyes parameter", q^2/b^4 (q is the electronic charge, and b is the bond length) (Phillips, 1973). This parameter is just the chemical hardness per unit volume (normalized). It also comes from "zero-order" density functional theory (the Heisenberg Principle) as applied to metals (Gilman, 1971).

Shear stiffness is considerably more difficult to relate to other properties than bulk stiffness because it depends on the stiffnesses of particular bond angles. Thus it cannot be described in terms of electron densities (or any other scalar quantity) alone. The bulk moduli of homopolar covalent crystals (Group IV) correlate linearly with their chemical hardnesses, but the C_{44} shear moduli do not, the shear stiffness of diamond being about 35% too high. This high value of shear stiffness for diamond was rationalized in Chapter 13 in terms of the bimodal electron-density distributions along the C–C bonds.

15.3 Plastic resistance (physical hardness)

The most common scales of physical hardness are the Moh scale used by mineralogists, and the Vickers scale used by mechanical engineers. The former measures scratch resistance, while the latter measures indentation resistance. Scratch resistance is a combination of resistance to fracturing, resistance to flowing, and resistance to phase transformation. It has no direct connection to either the bulk modulus or to chemical hardness (as presently defined). That is why the correlation plots of Yang, Parr, and Uytterhoeven (1987) scatter markedly.

Indentation resistance is a combination of resistance to plastic flow, phase transformation, and fracturing. When flow dominates, experiments have verified that dislocation motion is the microscopic transport process that generates the shape change (indentation). The phenomenon that limits the process is dislocation mobility, and it has been shown that the activation energy for this is proportional to the band gap in Group IV crystals (Gilman, 1993a). As pointed out above, the band gap defines the chemical hardness, so plastic flow is related to chemical hardness.

15.4 Shear-induced phase transitions

At low temperatures (below the Debye temperature), compressed covalent crystals undergo structural transitions in which the bond lengths do not change significantly, but the bond angles do. The prototype is the change in silicon from the diamond structure to the β-tin structure (or zinc blende to diatomic β-tin). The topology of the chemical bonds is retained (i.e., the coordination number). The transformation is martensitic, no atomic diffusion being involved.

These transformations are usually attributed to an effect of pressure, but they require a large change in shape and symmetry (cubic to tetragonal) so it is clear that they are driven by shear strains (Gilman, 1993b). Although the volume shrinks during the structural change, this is a secondary effect caused by the symmetry change (analogous with the change in volume of an accordion at constant component length of the instrument).

Before the transformation the material is often an insulator (or semiconductor), and afterwards it is a metal; so the chemical hardness changes from some finite value to zero. According to the "principle of maximum hardness" the material usually returns to its more stable state when the applied forces are removed. Conversely, since the chemical hardness is a measure of the stability, it might be expected that the stress required to cause the transformation would increase with chemical hardness. This is indeed the case (Figure 15.1). In this figure, the chemical hardness values are taken to be one-half the minimum band gaps of the various crystals. The existence of such a correlation was first suggested by Jamieson (1963) at a time when only a few data were available.

15.5 Shear-induced chemical reactions

Although allotropic transformations are a kind of isomerization reaction in which bond geometry changes, they are different from conventional reactions in which bond species change. Most reactions are thermally or photonically activated, and the role of chemical

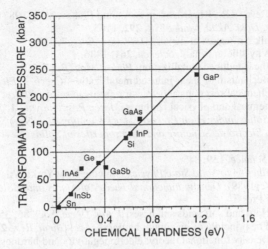

Figure 15.1 Critical values for pressure-induced phase transformations in a variety of covalent tetrahedral crystals versus chemical hardness.

hardness in these has been extensively discussed in the literature (Pearson, 1991). However, there is a large class of reactions that is athermal, being driven by mechanical potentials. These were first studied systematically more than 65 years ago by Bridgman (1935). His work was extended by Enikolopyan *et al.* (1987), Boldyrev (1986), Graham (1993), and others.

The author has proposed that the general mechanism of mechanochemical reactions is that elastic shear strains applied to molecules, or solid aggregates, cause an increase in the energies of their bonding states, and a decrease in the energies of their anti-bonding states (Gilman, 1995a). This decreases the size of the gap in the bonding energy spectrum, thereby destabilizing (decreasing the chemical hardness) of the system. When the hardness vanishes, the bonding electrons become delocalized because they can be in anti-bonding states without expending any energy, Thus, the material becomes very reactive. Another way of saying this is that the activation energy vanishes (Gilman, 1995b). An example of such a reaction is the thermite reaction: $Fe_2O_3 + 2Al \rightarrow Al_2O_3 + 2Fe$. This can be thermally activated, but it can also be driven by elastic shear strain at low temperatures.

References

Boldyrev, V. (1986). Mechanical activation of solids and its application to technology, *J. Chim. Phys.*, **83**, 821.

Bridgman, P.W. (1935). Effects of high shearing stress combined with high hydrostatic pressure, *Phys. Rev.*, **48**, 825.

Brinck, T., Murray, J.S., and Politzer, P. (1993). Polarizability and volume, *J. Chem. Phys.*, **98**, 4305.

Burdett, J.K. (1995). *Chemical Bonding in Solids*, p. 298. Oxford: Oxford University Press.

Edwards, P.P. and Rao, C.N.R. (Editors) (1995). *Metal–Insulator Transitions Revisited*. London: Taylor & Francis.

Enikolopyan, N.S., Vol'eva, V.B., Khzardzhyan, A.A., and Ershov, V.V. (1987). Explosive chemical reactions in solids, *Dokl. Akad. Nauk SSSR*, **292**, 1165.

Gilman, J.J. (1971). Bulk stiffnesses of metals, *Mater. Sci. Eng.*, **7**, 357.

Gilman, J.J. (1993a). Why silicon is hard, *Science*, **261**, 1436.

Gilman, J.J. (1993b). Shear-induced metallization, *Philos. Mag. B*, **67**, 207.

Gilman, J.J. (1995a). Mechanism of shear-induced metallization, *Czech. J. Phys.*, **45**, 913.

Gilman, J.J. (1995b). Chemical reactions at detonation fronts in solids, *Philos. Mag. B*, **71**, 1057.

Gilman, J.J. (1997). Chemical and physical hardness, *Mater. Res. Innovat.*, **1**, 71.

Graham, R.A. (1993). *Solids under High-Pressure Shock Compression*. New York: Springer-Verlag.

Harrison, W.A. (1989). *Electronic Structure and the Properties of Solids*. New York: Dover Publications.

Jamieson, J.C. (1963). *Science*, **139**, 845.

Kittel, C. (1976). *Introduction to Solid State Physics*, 5th edn. New York: Wiley.

Parr, R.G. and Yang, W. (1989). *Density Functional Theory of Atoms and Molecules*. New York: Oxford University Press.

Pearson, R.G. (1963). Hard and soft acids and bases, *J. Am. Chem. Soc.*, **85**, 3533.

Pearson, R.G. (1993). The principle of maximum hardness, *Acc. Chem. Res.*, **26**, 250.

Pearson, R.G. (1991). Density functional theory: electronegativity and hardness, *Chemtracts Inorg. Chem.*, **3**, 317.

Pearson, R.G. (1997). *Chemical Hardness*. New York: Wiley-VCH.

Phillips, J.C. (1973). *Bonds and Bands in Semiconductors*. New York: Academic Press.

Rasky J.D. and Milstein, F. (1986). Pseudopotential theoretical study of the alkali metals under arbitrary pressure: density, bulk modulus, and shear moduli, *Phys. Rev. B*, **33**, 2765.

Rose, J.H., Smith, J.R., and Ferrante, J. (1983). Universal features of bonding in metals, *Phys. Rev. B*, **28** (4), 1835.

Yang, W., Parr, R.G., and Uytterhoeven, L. (1987). New relation between hardness and compressibility of minerals, *Phys. Chem. Miner.*, **15**, 191.

Section V

Plastic strength

16

Macroscopic plastic deformation

Some solids are very hard and brittle, while others are soft and ductile. These differing behaviors are related to differences in resistance to plastic flow. Continuum mechanics cannot account for the differences. A quantum theory is required. In the case of plastic flow there are two levels of quantization that must be considered. The first is that, unlike elasticity, plastic flow is not a homogeneous process. It requires the inhomogeneous creation and propagation of dislocation lines. The displacements at these lines are quantized. The second is that dislocation lines themselves do not move homogeneously, except in simple metals. Kinks form on the lines. The displacements at these kinks localize and quantize the mobility process. Therefore, the wave mechanics of the bonding electrons determines the kink mobilities which in turn determine hardness and softness. Only by combining classical and quantum mechanics can a complete solid mechanics theory be developed.

Strong structures fail either because elastic deflections become too large in them, or because plastic flow occurs leading to large plastic deflections, or worse to fracture. Thus an understanding of the nature of plastic resistance is important for the design of strong structures. The subject is very complex because plastic flow is so very heterogeneous, at virtually every one of the 12 levels of aggregation from atoms to large engineering structures. At each level the heterogeneity manifests itself in different ways, all of which tend to be complicated. Therefore, no attempt will be made here to discuss the subject as a whole. Only a small part will be considered, namely, the intrinsic sources of plastic resistance. That is, those that are directly related to the fundamental chemical bonding within a material. The extrinsic sources of resistance that are associated with the various levels of aggregation will be mentioned but not considered in detail.

Elastic waves transport strains from places where tractions act on surfaces to other parts of elastic bodies. The strains are displacement gradients, and are reversible (if done isothermally). The work that is done to create them is stored in a material as strain energy.

Elastic deformation is usually temporary, well ordered, and mostly affine. Plastic deformation is usually permanent, chaotic, and anything but affine. Since the work needed to cause elastic deformation is stored in the deformed material in the form of changes of electronic structure (strain energy), it can be quickly recovered almost completely. Also, elastic deformation can propagate rapidly from one position to another. In contrast, the work that causes plastic deformation is dissipated as heat, as well as configurational entropy, so almost none of it can be recovered, and the deformation can propagate only relatively slowly.

An immediate conclusion is that elastic strains are state variables that depend only on the values of other state variables ($\pm$ pressure, shear stresses, temperature, electromagnetic fields, gravitational fields, etc.). Whenever these are given a particular set of values, the elastic strains (relative to a standard state) acquire corresponding values. Thus the state variables are connected in a definite way by an equation of state, and each specific state has a corresponding set of state variables.

In passing from one state to another an elastic material may lose, or gain, small amounts of energy due to anelastic effects which convert small amounts of elastic energy into entropy and thus are not recoverable. These losses may be associated with specific mechanisms, and their amounts can be calculated quantitatively (at least in principle).

In contrast to the well behaved case of elastic deformation, there is little about plastic deformation that is not chaotic, starting with the distribution of displacements that is associated with it.

For centuries it was thought that during plastic deformation the distribution of plastic displacements (the plastic displacement field) was microscopically uniform. Until the acceptance of the atomic theory of matter at the end of the nineteenth century, there was no reason to think otherwise. Localization of the deformation into shear bands was well known, of course, but it was not realized that this fragmentation continued beyond microscopic dimensions for 3–4 orders of magnitude further, down to atomic dimensions. Furthermore, the instrumentation that recorded "stress–strain curves" drew predominantly smooth lines with little change in passing from the elastic to the plastic regime (sometimes serrated curves were observed, but they were not the norm).

It is now known that plastic deformation is anything but uniform. It is heterogeneous all the way down to the atomic level and somewhat beyond. The somewhat beyond refers to the fact that the quantum mechanical amplitude functions are disturbed in a non-uniform way at the kinks on dislocation lines. The contrast with the homogeneous nature of elastic deformation is striking.

Figure 16.1 illustrates the discontinuity in behavior that occurs at the "yield point" of an elastic plastic material. The continuity of the elastic stress–strain line is lost. The steps in the figure can be quite large, or they can be exceedingly small. A limit is reached when a kink on a dislocation line moves one atomic distance, b. Then, in an atomic volume, say b^3, the plastic shear deformation is approximately unity. If this is averaged over a macroscopic volume of a cubic centimeter, the macroscopic plastic deformation is about 10^{-23}. In practice, this is much too small to be observed, but it makes the point that plastic yielding is an exceedingly discontinuous process.

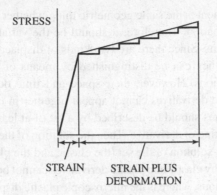

Figure 16.1 Comparison of elastic and plastic parts of a stress–strain curve. The elastic part is continuous, while the plastic part is discontinuous on a fine scale.

16.1 Distinction between elastic and plastic deformations

Small changes of the shapes of solid bodies can be conveniently described in terms of displacement gradients, that is, in terms of a field of displacements normalized by dividing them by a corresponding field of local gages, yielding a field of displacement gradients. To provide a useful description these fields must be continuous and compatible. Furthermore, in the elastic case, they must be differentiable with respect to both time and space. Plastic displacement gradients do not meet this last criterion because they are irreversible, and they are microscopically discontinuous.

Unfortunately, for historical reasons, the word "strain" has been used to describe both elastic and plastic deformations, sometimes differentiated by subscripts, but their physical bases are fundamentally different. Therefore, their names should be distinct. Also, it is improper to add, subtract, multiply, or divide them. In this book, the word "strain", and its corresponding symbol, ϵ, will be restricted to the elastic case. For the plastic case, the word "deformation", and its corresponding symbol, δ, will be used.

It is especially important to recognize that quantities derived from ϵ and δ are also physically distinct. In particular, $(d\sigma/d\epsilon)$, an elastic modulus, is physically very different from $(d\sigma/d\delta)$ which is not (repeat not) a modulus. The latter relates to the dissipation of stress, but not to its propagation. The slope $(d\sigma/d\delta)$ may legitimately be taken to be a strain-hardening coefficient, however. A formal distinction is that $(d\sigma/d\delta)$ may be either positive or negative, whereas $(d\sigma/d\epsilon)$ must always be positive.

Negative values of $(d\sigma/d\delta)$ are important because they produce a mode of plastic insta-bility that results in localized plastic shear bands. The negative values are created whenever the rate of multiplication of dislocation lines increases the deformation rate more rapidly than is necessary to keep up with the deformation rate that the external system is trying to impose. The external system may be the loading machine, the strained material around the tip of a crack, etc.

Physically, a displacement is the same geometric thing whether it results from an elastic strain or a plastic deformation. Therefore, it should be the variable of choice in dealing with elastoplastic problems. Since there are two kinds of displacement, although they are geometrically identical, they can be distinguished by means of subcripts; and they can appear together in equations. However, their space and time derivatives have different physical meanings, so the derivatives cannot appear together in the same equation. Thus most elastoplastic problems should be described by a set of at least two equations each of which contains only one kind of derivative. Then the solution of the corresponding problem requires the simultaneous solution of the set (the elastic and the plastic equations).

An additional reason why elastic and plastic derivatives cannot be mixed is that the elastic displacements are continuous, whereas the microscopic plastic displacements are quantized in units of the Burgers displacement, b. To some extent these quantized displacements can be smoothed out in macroscopic problems, but this does not eliminate the other differences between the elastic and plastic displacements, i.e., the fact that one is conservative, while the other is non-conservative.

The second derivative of an elastic displacement with respect to position is an elastic modulus (either bulk or shear). This is not the case for the second derivative of the plastic displacement since the latter is not conserved. Thus there is no "plastic modulus". This is important in interpreting elastic plastic impacts, plastic zones at crack tips, local yielding, upper yield points, etc.

16.2 Plastic equation of state

If a general mechanical equation of state existed, it would be possible to approach any given point in a stress, σ, deformation, δ, temperature, T, space along an arbitrary path, and the state of the material at the point would always be the same. In other words the material would obey an equation analogous to the state equation for gases (A is a constant):

$$PV = AT \qquad (16.1)$$

This has been extensively tested by comparing its partial derivatives at various plastic states (P, V, T points). It is obeyed as long as a material is elastic, but as soon as the yield point is reached, it fails to give consistent values. Equations of this type were extensively tested in the 1930–1950 time period. Some of the results are discussed by Tietz and Dorn (1949).

Another type of state equation is one in which the rate of a reaction depends only on time and temperature if the stress, or the deformation, is held constant. This is true for simple chemical reactions in dilute systems. If it is also true for mechanical systems, the plastic behavior should depend only on a combination of time, t, and temperature, T, known as the Zener–Hollomon parameter (1947):

$$\text{rate} \sim t[\exp(-Q/kT)] \qquad (16.2)$$

Accordingly, deformation at a high rate and a low temperature should be the same as deformation at a low rate and a high temperature. To a considerable extent this is true, but the substitution of time for temperature is not quantitative. Therefore, this parameter is useful for making interpolations, but it is not reliable for extrapolations.

16.3 Modes of plastic deformation

Although a majority of plastic deformation results from linear shear, like the shearing of a deck of playing cards, there are several other modes. In the playing card analogy, the amount of shearing at each position through the thickness of the deck varies in a random fashion, but the shear at a given position in a crystal is a multiple of a translational gliding element of the crystal structure.

A variant of the first mode (called translation-gliding) is twin-gliding (or twinning) in which a crystal structure shears into a mirror image of itself. This requires the same amount of shear to occur at each atomic level of a suitably oriented crystal structure (Figure 16.2).

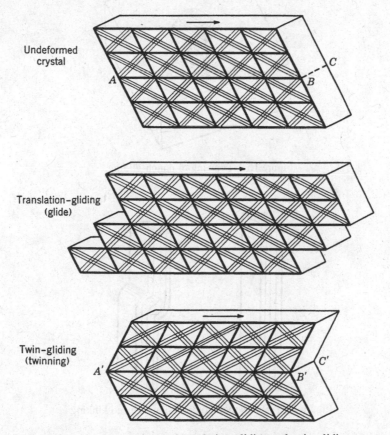

Figure 16.2 Comparison of translation-gliding and twin-gliding.

Plastic strength

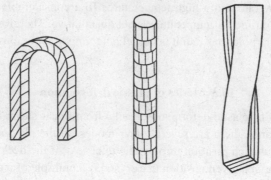

Figure 16.3 Schematic drawings of bend-, rotation-, and bend-twist-gliding.

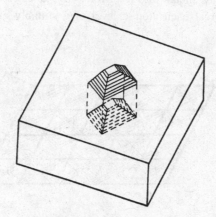

Figure 16.4 Drawing of punch-gliding.

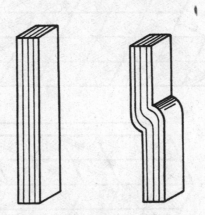

Figure 16.5 Diagram of compression-induced kinking.

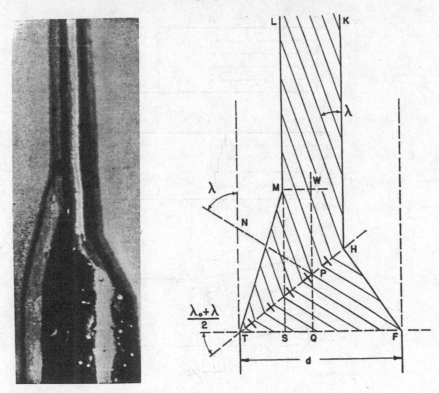

Figure 16.6 Accommodation kinking at the end of a tensile specimen.

A deck of cards may also be bent (bend-gliding), or rotated about a normal to the plane of each card (rotation-gliding), or deformed by a combination of bending and twisting (twist-gliding), see Figure 16.3.

Using a set of two planes for sliding of their structures, some crystals such as those with the cesium chloride structure may be plastically "punched" (punch-gliding), Figure 16.4.

Finally, if the macro-deformation is non-uniform, various forms of kinking may occur. Figure 16.5 illustrates compression kinking. This provides a means for a crystal to accommodate inhomogeneous deformation without changing its density (Figure 16.6). If the inhomogeneous deformation cannot be accommodated, fracture ensues. Figure 16.7 shows how a tensile strain, $\epsilon \approx \gamma/4$ where γ is the shear strain, can be generated.

Kinking has a special compatibility condition. The materials on the two sides of a kink plane must have matching heights. This can only be the case if the kink plane bisects the crystallographic angle between the orientations on the two sides.

Gliding on multiple planes, or in multiple directions, is another means for accommodating inhomogeneous deformation. If the deformation does not occur with equal

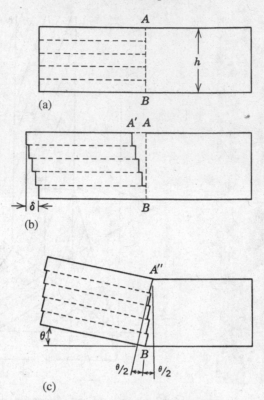

Figure 16.7 The lamellar nature of translation-gliding requires that a shear discontinuity lie in a plane that bisects the misorientation angle between the material on the two sides of the discontinuity, AB. (a) Undeformed state; (b) material on the left sheared, δ/h, note that A'B' $\neq$ AB; (c) material on the left joined to that on the right along a matched boundary, A''B.

ease on various planes, or in various directions, a crystal is said to be plastically anisotropic.

For photographs of the various modes of plastic deformation in actual crystals see Gilman (1969).

References

Gilman, J.J. (1969). *Micromechanics of Flow in Solids*, Chapter 3. New York: McGraw-Hill.
Hollomon, J.H. and Jaffe, L.D. (1947). *Ferrous Metallurgical Design*. New York: Wiley.
Tietz, T.E. and Dorn, J. (1949). The effect of strain histories on the work hardening of metals. In *Cold Working of Metals*, p. 163. Cleveland, OH: American Society for Metals.

17

Microscopic plastic deformation

Plastic deformation is transported by dislocation lines which have discrete vectorial displacements associated with them (Figure 17.1). The deformation is irreversible. The work that is done to create it is converted into heat, or stored in the material as defects (mostly dislocation dipoles).

17.1 Plasticity as linear transport

Orowan (1934) proposed a transport equation relating the *microscopic* behavior of the dislocation lines (N is the line flux, number/cm^2), and the *macroscopic* plastic deformation rate, $d\delta/dt$, of crystals. This relationship (t is time) is:

$$d\delta/dt = bNv \qquad (17.1)$$

This is similar to other transport equations, but not quite the same because dislocations are *lines* rather than quasi-*particles*. In the case of plasticity it is displacement (shape) that is transported. In all transport cases (plasticity, thermal conduction, electrical conduction, and diffusion) the form of the equation that connects the macroscopic observable flux with the microscopic events is the same. The right-hand side of Equation (17.1) consists of the product of the entity being transported (in this case, displacement), the concentration of carriers (in this case, the flux of dislocation lines), and the average velocity of the carriers (dislocation lines). For electricity, the right-hand side would be the product of the charge/carrier, the carrier concentration, and the average velocity, while the left-hand side would be the current.

The nature of δ requires some explanation because it is often confused with elastic shear strain (Figure 17.2). Although this explanation is partly a repetition of comments in Chapter 16 it is important to emphasize that δ is quite physically distinct from ϵ. Although δ has the same dimensions as elastic strain (displacement/length), there is no strain energy associated with it. Therefore, it is not reversible, and it does not obey an equation of state. Most of the work used to create the plastic deformation, δ, is converted into heat (i.e., dissipated), and the remainder is converted into various material defects through stochastic processes.

193

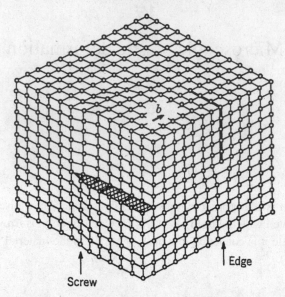

Figure 17.1 Dislocated simple cubic crystal. The top lower quadrant has glided over the bottom lower quadrant along the mid-height plane. The boundary line between the glided part of the mid-height plane and the unglided part is the dislocation line. Where this boundary line is parallel to the displacement vector **b** it has a "screw" configuration. Where the boundary is perpendicular to **b** it has an "edge" configuration. Note that the structure is restored behind the dislocation line.

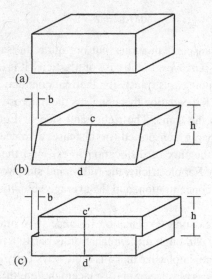

Figure 17.2 Comparative characteristics of elastic and plastic shear deformation. (a) Undeformed block of material. (b) Block elastically deformed, shear strain $\epsilon = b/h$, the line cd is stretched. Strain is reversible. (c) Block plastically deformed, shear defomation $\delta = b/h$, the line c'd' is not stretched. Deformation is not reversible.

Another important distinction between elastic and plastic strains in crystals is that the former can be infinitesimal, whereas the latter must be finite because crystals have discrete atomic structures. At the macroscopic level, δ can be very small, albeit finite. At the atomic level, it cannot be less than about one.

Consider an atomic length, l, of dislocation line (in atomic units, $l = 1$). Assume the crystal has a simple square structure with a lattice spacing of 1. Then the height h of the glide plane is 1, $b = 1$, and $\delta = \Delta u/h = b/h = 1$ which is clearly finite, and which is quantized in units of b. This contrasts markedly with a local elastic shear strain ($\mathrm{d}u/\mathrm{d}y$) which must be less than about $1/100$.

Experiments have confirmed the validity of Orowan's Equation (17.1) quantitatively (Gilman and Johnston, 1962). It has been generalized (for a single glide system) to the form (Gillis and Gilman, 1965):

$$\mathrm{d}\delta/\mathrm{d}t = gb[\text{line integral } (\mathbf{n_o} \cdot \mathbf{v})\,\mathrm{d}l]_l \qquad (17.2)$$

where l is the length in a unit volume of a set of closed loops with unit outer-normal vectors, $\mathbf{n_o}$, g is a geometric factor of order unity, and b is the magnitude of the displacement across the dislocation line. For meandering lines (open loops), $\mathbf{n_o}$ must be replaced by $\mathbf{n_1}$ which is written in terms of a unit tangent vector, $\mathbf{s}$, as follows: $\mathbf{n_1} = \mathbf{s} \times (\mathbf{b} \times \mathbf{s})/|\mathbf{b} \times \mathbf{s}|$. For any particular material, b is a constant.

Since the macroscopic deformation rate is proportional to the product of the concentration of carriers (dislocations), and their average velocities, a macroscopic measurement always has an ambiguous microscopic interpretation. For a given value of the product, the concentration may be high while the velocity is low, or vice versa. Also, this ambiguity provides various irreversible paths to reach a particular plastic state, so no equation of state exists.

The *average* dislocation velocity always corresponds to the instantaneous value of the stress. That is, dislocations do not "run away"; their average velocities are steady at constant stress, although the velocities fluctuate from point to point.

17.2 Multiplication of dislocations

Dislocation multiplication at the microscopic level obeys first-order kinetics. This can be deduced quite simply from the kinematics. That is, from the Orowan equation together with the observation that the dislocation density increases linearly with the plastic deformation at constant stress (constant velocity). Replacing N by L, which is the length of line per unit volume:

$$L = \alpha\delta \qquad (17.3)$$

where α is a constant, and the small initial dislocation density is neglected. An increment of plastic deformation during a time interval $\mathrm{d}t$ is thus given by:

$$\mathrm{d}\delta = (bLv)\,\mathrm{d}t \qquad (17.4)$$

Then, letting A be the constant factors $bv\alpha$, and using Equation (17.1):

$$dL/dt = AL \tag{17.5}$$

showing that the kinetics is first order. Upon integration:

$$L = L_0 \exp(At) \tag{17.6}$$

so the growth occurs exponentially.

The process that yields the first-order kinetics of Equation (17.5) was first clearly described by Koehler (1952), and was experimentally confirmed in detail by Johnston and Gilman (1960). It comes from the symmetry of screw dislocations. In an elastically isotropic medium, screw dislocations are cylindrically symmetric (except at their cores which may have lower symmetry). Therefore, they can move in any direction normal to their line tangent with nearly equal ease. The plane of maximum shear stress will be favored, but there are always two of these. Also, the crystallographic symmetry around the Burgers vector, and the extension of the core (or lack of it) may bias the direction of motion. The net result is that there is often little to restrain dislocations from wandering from one glide plane to another, and then back to a plane parallel to the original one (Figure 17.3). The way in which internal stresses cause this multiple-cross-gliding is described in the theory of Li (1961). This theory yields increasing rates of cross-gliding with increasing stress. A stochastic mechanism for the Koehler process has been described by Gilman (1997).

Through multiple-cross-glide a dislocation line can undergo "random walk" onto a parallel neighboring plane. Each time this occurs two jogs are created. They lie perpendicular to the Burgers vector, so they can move parallel, but not perpendicular, to it. Thus they produce drag on the screw segments. The dipoles from the non-crossover case can only be eliminated by climb. This is diffusion controlled, so at low temperatures dipoles accumulate and impede the motions of other dislocations, thereby causing strain-hardening.

In principle, a highly deformed macroscopic crystal may contain just one dislocation line (including its virtual connectors at surfaces). This one line will have been derived from just one initial critical nucleus. The length of such a line in a one centimeter cube of material might be as much as 10^8–10^9 km.

A general feature of the multiplication process is that the rate is very stress (or velocity) dependent. It has been shown experimentally that a dislocation loop can be expanded slowly (under a small applied stress) without causing it to multiply, but if it is expanded quickly (under a higher stress) it will multiply profusely. This is one of the fundamental reasons why there is no plastic equation of state in terms of the macroscopic variables.

Another corollary of the Koehler process is that the dipole debris gradually reaches high concentrations which causes more or less degradation of the strength. The effect depends on the type of chemical bonding in the material.

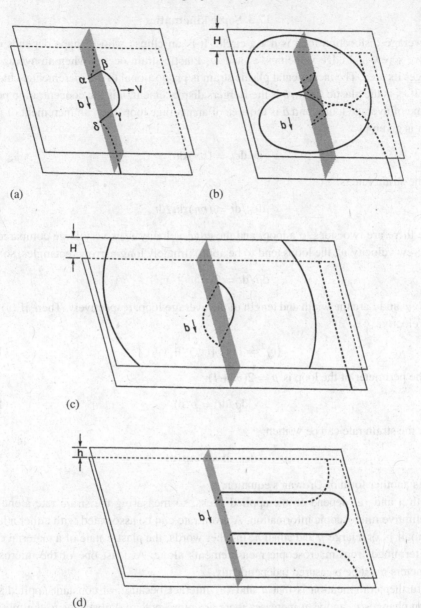

(a)

(b)

(c)

(d)

Figure 17.3 Koehler multiple-cross-glide process. (a) A screw dislocation line moving on the bottom of two glide planes spaced H apart, cross-glides onto an inclined (shaded) plane, and then cross-glides again back onto the top plane. Segments $\alpha\beta$ and $\delta\gamma$ are perpendicular to **b**, so they cannot move in the forward direction. Thus a loop begins to form on the top plane. (b) The loop on the top plane spreads, crossing over loop segments on the bottom plane, and meets itself on the top plane to the left of the inclined plane. (c) The original line is moving forward on the bottom plane, and two lines lie on the top plane (one moving forward and the other backward) and another double cross-glide event has just occurred at the inclined plane. (d) Alternative case when the spacing of the planes is $h < H$ such that lines cannot pass over one another. Then two edge-dipoles are left behind as debris.

17.3 Some kinematics

An average dislocation loop is not a circle. It is an ellipse whose shape depends on the relative screw and edge velocities, v_s and v_e. Plastic strain occurs when an average loop changes its area. The incremental plastic strain is proportional to the incremental change in area. If ϵ_p is the plastic strain, b is the Burgers displacement, n is the concentration per unit volume of average loops, and a is the area of an average loop, then an increment of plastic strain is given by:

$$d\epsilon_p = (bn)\,da \tag{17.7}$$

and the strain rate is:

$$d\epsilon_p/dt = (bn)\,da/dt \tag{17.8}$$

Since there are two sides to a loop, and the edge velocity v_e is very large compared with the screw velocity v_s, the loops tend to be approximately long narrow rectangles, so:

$$da/dt = 2(v_e w + v_s l) \tag{17.9}$$

where w and l are the width and length of the average loop, respectively. Then, if $\langle v \rangle$ is the rms velocity:

$$\langle v \rangle^2 = (1/2)[(v_e)^2 + (v_s)^2] \tag{17.10}$$

and the perimeter of the loop is $p = 2(w + l)$:

$$da/dt = p\langle v \rangle \tag{17.11}$$

Thus, the strain rate can be written:

$$d\epsilon_p/dt = bnp\langle v \rangle \tag{17.12}$$

This is another form of Orowan's equation.

Both n and $\langle v \rangle$ depend on the applied stress, so measuring the strain rate alone gives no definitive microscopic information. A given rate can be associated with either a large n and small p, or a large p and small n. In other words, the plastic state of a material cannot be determined from macroscopic measurements alone. At least one of the microscopic parameters must be measured independently.

A further complication is that n and $\langle v \rangle$ interact because, at constant applied stress, motion changes n, and as n increases there are more potential obstacles to the motion of the lines and this tends to reduce $\langle v \rangle$. In addition, as n increases, an increasing fraction of the loops becomes immobilized. The remaining fraction of the loops that is mobile can be written:

$$n_m = n(e^{-\delta n}) \tag{17.13}$$

where δ is an attrition coefficient, and the term in parentheses decreases as the total concentration of loops increases.

At any constant strain rate, there is no dL/dt term. However, if the strain rate is changing, we have the following:

$$d^2\epsilon/dt^2 = (d\epsilon/dt)[(1/v)(dv/dt) + (1/L)(dL/dt)] \qquad (17.14)$$

If the velocity is very high, approaching the speed of sound, dv/dt becomes very small so the first term drops out, but the second term which describes multiplication remains. It may be written:

$$d(\ln L)/dt = d(\ln n)/dt + d(\ln p)/dt \qquad (17.15)$$

where the first term on the right describes the nucleation of new loops (plus any annihilation that occurs), and the second describes the growth of the perimeters of existing loops (plus any shrinkage that occurs). Both terms require dislocation motion, so good mobility is a prerequisite for rapid transient response, and hence for ductility. This disqualifies silicon for example. Note that both terms describe first-order kinetics so the total line length grows exponentially in time.

When the applied displacement rate exceeds the response rate of the specimen, the stress rises, and inversely the stress falls. If the two rates are equal the stress remains constant (at the upper yield point).

17.4 Importance of dislocation mobility

Both initial plastic deformation rates, and increases in rates resulting from multiplication, depend on the ability of dislocations to move. Therefore, dislocation mobility is a key parameter. So the next and longest chapter in this book will be devoted to this subject.

In simple metals (non-transition), and highly ionic crystals, dislocation mobilities are very high because no "chemistry" occurs when the cores of dislocations in these substances move. However, when the core of a dislocation in a covalent substance moves (crystalline or otherwise), the local chemical structure (i.e., the local electronic structure) must change to a transition structure, and then change back to the initial structure. Such reactions change the symmetry locally (in particular, the bonding structure), and this leads to a large barrier to dislocation motion.

The rest of the discussion of mobility will be concerned with $\langle v \rangle$, and its components v_e and v_s. The flow stress level is proportional to these velocities because they are usually steady-state velocities and, since plastic flow is a dissipative process, the energy losses associated with the motion just balance the input work done by the applied stress. In other words, the macroscopic plastic resistance is largely determined by the microscopic resistances to the motion of individual dislocation lines.

References

Gillis, P.P. and Gilman, J.J. (1965). Dynamical dislocation theory of crystal plasticity. I. The yield stress, *J. Appl. Phys.*, **36** (11), 3370.

Gilman, J.J. and Johnston, W.G. (1962). Dislocations in lithium fluoride crystals. In *Solid State Physics*, ed. F. Seitz and D. Turnbull, Volume 13, p. 147. New York: Academic Press.

Gilman, J.J. (1997). Mechanism of the Koehler dislocation multiplication process, *Philos. Mag. A*, **76** (2), 329.

Johnston, W.G. and Gilman, J.J. (1960). *J. Appl. Phys.*, **31**, 632.

Koehler, J.S. (1952). *Phys. Rev.*, **86**, 52.

Li, J.C.M. (1961). *J. Appl. Phys.*, **32**, 593.

Orowan, E. (1934). *Z. Phys.*, **89**, 605, 614, 634.

18

Dislocation mobility

18.1 Introduction

Since dislocations "multiply" as they move (i.e., they increase their length), and since they are almost always present initially in structural materials, their most important property is their mobility. If this is very small, as it is in covalently bonded crystals like silicon at temperatures below the Debye temperature (about 920 K in Si), the material is brittle, tending to fracture before it deforms plastically. If the mobility is large, as it is in nearly perfect gold, or copper, crystals, then the material is very ductile, or malleable. Such crystals may be hammered violently without fracturing them.

The range of dislocation mobilities is very large as measured by the stress needed to move a dislocation at an observable rate in a laboratory experiment. This range starts at zero in a nearly perfect metal, and ends at about $G/4\pi$ in a covalent crystal, where G is the appropriate shear modulus (G of the (111) plane is about 5 Mbar for diamond). At the high end of this range, the velocity is proportional to the applied stress so the mobility is liquid like, and the material has no intrinsic barrier to dislocation motion (if finite barriers are observed, i.e., finite yield stresses, they result from extrinsic factors). At the low end of the range, the mobility is intrinsic, determined by the chemical structure of the crystal.

The extrinsic factors that create finite yield stresses are numerous. They include: vacant lattice sites; interstitial atoms; impurities of the wrong size, or valence; color centers; other parallel, or perpendicular, dislocations; grain boundaries; anti-phase boundaries; twins; precipitates; free surfaces; and more.

18.2 Mobilities, general

Until the late 1920s, it was not at all understood how the wide range of mechanical activity described above could occur for solid materials. The search had been for some kind of concerted mechanism that would allow solids to change their shapes uniformly (homogeneously). Then, several people independently realized that a disconcerted mechanism might account for what was being observed, that is, non-uniform (inhomogeneous) flow. The overall plastic flow might start with localized internal sliding surrounded by a line bounding the slided area of a plane from the unslided part. This bounding line

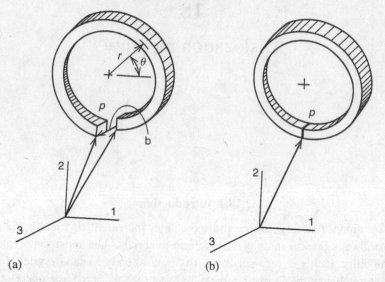

Figure 18.1 Typical elastic dislocation: (a) cut and bent steel ring, (b) ring flattened, ends welded together, and applied forces removed, yielding a ring in a state of self-equilibrated stress.

came to be known as a "dislocation line" because the configuration was quite similar to the self-equilibrated strained bodies of the theory of elasticity (Figure 18.1). The word *dislocation* was coined by Love (1944), for the elastic theory developed by Volterra (1907).

It was immediately realized that a local dislocation line could slide along its plane much more easily than the whole plane could slide concertedly (Orowan, 1934; Polanyi, 1934; Taylor, 1934). This is analogous with the fact that it is much easier to move a ruck in a rug than it is to move the whole rug concertedly. Furthermore, it is much easier to move a ruck if the rug lies on a polished surface than if the surface is rough. Thus rucks have various mobilities, and so do dislocation lines.

The reason for the high mobility is shown schematically in Figure 18.2 where the horizontal axis is exaggerated in order to indicate how the forces on either side of the central location are balanced. The balancing makes the net force very small, and nearly independent of the position of the center (core) of the dislocation.

In fact, if the mobility, μ_o, is defined as the velocity, v, created by a given driving force (shear stress τ times unit of displacement, b), then:

$$v = \mu_o b \tau \tag{18.1}$$

or, since for most cases of practical interest the mobility is not a constant, but depends on the stress:

$$\mathrm{d}v = \mu(\tau) b \, \mathrm{d}\tau \tag{18.2}$$

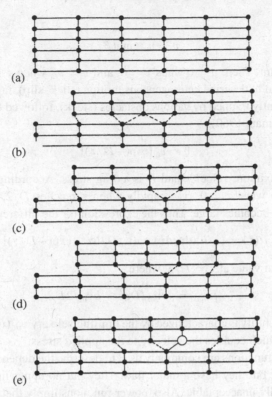

(a)

(b)

(c)

(d)

(e)

Figure 18.2 Schematic drawing indicating how the forces on the atoms at the center of a dislocation change as it moves: (a) initial state, no dislocation; (b) symmetric position, forces balanced, no net force; (c) another symmetric position, forces balanced, no net force; (d) asymmetric position, forces unbalanced, small net force; (e) indicating that an impurity atom unbalances the forces, yielding a net force.

That is:

$$\mu(\tau) = (1/b)(dv/d\tau) \tag{18.3}$$

and it may be seen that μ is the inverse of a viscosity, η. At low temperatures in metals, the minimum viscosity for dislocation motion, caused by the Fermi electron liquid, is about 10^{-4} dyn s/cm^2 (poise). Thus, in pure, simple, metals, the maximum mobility is about 10^4 cm^2/dyn s; and it ranges downward to infinitesimal values in materials like rocks, and cutting tools, for ordinary applied stress levels. It also becomes small at high stresses where the viscous drag stress approaches $G/4\pi$ as the dislocation velocity approaches the velocity of shear waves ($\sqrt{G/\rho}$ is the square root of the shear modulus divided by the mass density) in the material.

Two algebraic functions that describe zero velocities at zero stresses and yield limiting velocities at high stresses are the hyperbolic tangent, and the reciprocal exponential. The first can be used to describe Newtonian viscosity (stress proportional to deformation rate at low stresses):

$$v = v_m \tanh(\beta\tau) \qquad (18.4)$$

where β is a constant. Then, $v = 0$ when $\tau = 0$; and $v = v_m$ when τ is large. The second function can be used to describe heterogeneous motion (stick–slip), in which a dislocation becomes intermittently trapped by various obstacles (stuck), followed by free runs (slip) at high velocities (Gilman, 1960):

$$v = v_m[\exp(-D/\tau)] \qquad (18.5)$$

where v_m is the maximum velocity, and D is a drag stress. According to this expression, the velocity is zero when $\tau = 0$, rises rapidly when τ reaches $D/2$, and asymptotically approaches v_m as τ becomes large. Thus the expression for the differential mobility is:

$$\mu(\tau) = (1/b)(dv/d\tau) = (v_m D/b\tau^2)[\exp(-D/\tau)] \qquad (18.6)$$

This has a maximum value at $\tau = D/2$, where:

$$\mu_{max} = 4bv_m/De^2 \approx bv_m/2D \qquad (18.7)$$

so the maximum mobility is characterized by the limiting velocity v_m ($v_m \approx (G/\rho)^{1/2}$), and the drag stress D which couples the velocity to the applied stress.

Empirical power functions are commonly used to describe the dependence of dislocation velocities on stress, but they have a major flaw. They put no upper limit on the velocity, so they are physically unacceptable. Also, power functions imply that the variation of the velocity about its average value is small. In fact, for the cases being described, the variation is very large because the motion is of the "stick–slip" type. Sticking is caused by extrinsic factors that create obstacles to dislocation motion. The fast "slipping" is limited by intrinsic viscous drag caused by the interactions of dislocations with conduction electrons, and density fluctuations (phonons).

18.3 Dislocations with low mobilities

In the 1940s, and 1950s, with the rise of interest in semiconductors, it was found that even the most pure Ge and Si crystals were brittle at temperatures below their Debye temperatures even though they contained dislocations. Why? Some researchers put their faith in a mechanical model, called the Peierls (1940) model. It combined a local set of sinusoidal forces at the glide plane of an edge dislocation with the forces of its elastic field (Cottrell, 1953). Peierls found a simple expression for the displacements along the glide plane that satisfies static equilibrium for this case. This is:

$$u(x) = -(b/2\pi)\tan^{-1}(x/w) \qquad (18.8)$$

where w is a width parameter, $w = a/2(1 - v)$, with a the glide plane spacing, and v Poisson's ratio. Using a local periodic potential with an adjustable shape instead of Peierl's sinusoid, Foreman, Jaswon, and Wood (1951) concluded that the best expression for the equilibrium width is:

$$w_{opt} = 2b/(1 - v) \approx 3w \text{ to } 4w \tag{18.9}$$

Since interatomic forces are very short range, this is too wide for physical consistency. Also, a sinusoidal potential is too smooth given that there is singularity at the center of a dislocation. Thus the Peierls theory is too simple to be physically consistent. A more consistent chemical theory will be presented later in this chapter.

In the Peierls model, the energy of a dislocation depends on its position relative to the crystal structure (a simple rectangular array of atoms). The maximum gradient of this energy function gives the stress, τ, needed to move the dislocation. Nabarro (1997) calculated this to be:

$$\tau = [2G/(1 - v)] \exp -(4\pi w/b) \tag{18.10}$$

where $3 < w/b < 4$. If $w/b = 3$, and $v = 1/3$, then $\tau/G = 2.4 \times 10^{-4}$. However, if $a = 5b/4$, so $w/b \approx 15\pi/4$, then $\tau/G = 2.3 \times 10^{-5}$ which is far too small for silicon, and too large to match observations in high purity metals. Therefore, the Peierls–Nabarro theory is not consistent with observations.

There is no single reason for dislocation mobility, or a lack thereof, because it depends strongly on the type of chemical bonding in a material. In the case of simple metals, the theoretical width of a dislocation is a few atomic spacings, but the bonding is very delocalized, so the energy of a dislocation is nearly independent of its position (Figure 18.2). However, the bonding in covalent crystals is highly localized to the regions between pairs of atoms (less than one atomic distance), and this depends strongly on the position of the center of a dislocation.

18.4 Steadiness of motion

Experiments indicate that dislocation motion is steady. That is, the work done to cause the motion is fully consumed during the motion. In other words the motion is fully dissipative, and critically damped. Observable "coasting" does not occur. The time used for acceleration and deceleration is quite short because the effective mass, m (per atomic length), is small, being approximately ρb^3, where ρ is the mass density. Since b is approximately the diameter of an atom, this is approximately one atomic mass.

Suppose a moderate stress of $10^{-4}G$ is applied, G being the shear modulus. Then the acceleration per atomic length is $10^{-4}Gb/m = 10^{-4}G/\rho b = 10^{-4}v_s^2/b$ where v_s is the shear wave speed. So, the time needed to go from zero to v_s is $10^4 b/v_s \approx 10^{-9}$ s, or one nanosecond. A result is that dislocations tend either to be not moving at all, or to be moving at "top speed", with little time lost in between.

It is probable that dislocations moving under the influence of an applied stress are subject to the fluctuation-dissipation theorem (Reif, 1965). Fluctuations are always present in solids because of the quantum mechanical zero-point vibrations. These vibrations are small, but always present. They interact with the dislocations just enough to keep them in place when there is no applied stress, if the material is free of other impediments. This accounts for the fact that the applied stress needed to cause measurable microscopic motion of dislocations is nearly the same as the stress needed to cause measurable macroscopic plastic deformation (Gilman and Johnston, 1962).

18.5 Resistance to individual dislocation motion

As mentioned previously, there are two broad classes of resistance to dislocation motion: intrinsic mechanisms and extrinsic mechanisms. So many sub-divisions exist for each of these, that only a few can be covered here. Therefore, the emphasis will be put on the intrinsic mechanisms for pure materials, after outlining the large number of extrinsic mechanisms. The latter tend to be complex, and therefore not directly related to electronic phenomena. They are discussed at length in textbooks on the mechanical properties of materials.

18.5.1 Extrinsic resistance

Extrinsic mechanisms are those in which dislocation lines as a whole interact with some structural feature of a material. The distinction between this and intrinsic resistance cannot be made rigorously, but nearly so. Extrinsic resistances will be simply listed with little discussion:

(1) *Internal stress fluctuations* – more velocity is lost when the net stress is below the average than is gained when it is above the average. Hence, stress fluctuations cause a virtual drag, or apparent dissipation.

(2) *Grain boundaries* – both the elastic field and the core structure of a dislocation are affected when it enters a grain boundary. This creates either an attraction or a repulsion. In most cases, the projections of unit lines lying normal to the glide planes on either side of the boundary are of different lengths at the boundary. Therefore, the configuration is plastically incompatible, and large local stresses arise. Only perfectly symmetric boundaries avoid this.

(3) *Twin boundaries* – similar to the grain boundary case.

(4) *Stacking faults, including anti-phase domain walls* – when they are sheared by dislocations passing through them, they increase their areas. Therefore, their energies increase, and they resist the passage of dislocations.

(5) *Kink bands*.

(6) *Dislocation multipoles, especially dipoles*.

(7) *Inclusions of foreign composition*.

(8) *Precipitates, including Guinier–Preston zones*.

(9) *Threading dislocations* – lines that thread through the primary glide plane.

(10) *Free surfaces*.

(11) *Cavities*.

(12) *Point defects* – including impurities, vacancies, and interstitials.

(13) *Domain walls* – ferroelectric and ferromagnetic.

(14) *Collective interactions* – like those that slow the flow of vehicular traffic.

18.5.2 *Intrinsic resistance*

When a dislocation line is at rest, a displacement (the magnitude of the Burgers vector) occurs across the dislocation line (Figure 18.3(a)). The gradient of this displacement (the shear strain) is concentrated at the dislocation's core (Figure 18.3(a)) with the maximum value of du/dx at the center. When the dislocation moves, the strain rate (or velocity gradient) distant from the dislocation is initially very small, but it rises to a maximum as the dislocation approaches, and then decreases back to a small value (Figure 18.3(b)) after the dislocation has passed. The maximum strain rate is given by (Gilman, 1968a):

$$(d\epsilon/dt)_{max} = (2b/\pi aw)v_d \tag{18.11}$$

or approximately v_d/a, where b is the Burgers displacement, a is the glide plane spacing, w is the core width, and v_d is the dislocation velocity. Note that this is the maximum for the entire strain field. The strain rate is a velocity gradient, and when multiplied by a viscosity coefficient gives a drag stress.

The viscosity coefficient measures the transport of momentum down the velocity gradient. This tends to reduce the gradient by moving faster material from the top region to the slower bottom region. Thus, the velocity gradient at a dislocation core is reduced by moving quasi-particles (electrons or phonons) from above to below the glide plane. Another way to reduce the gradient is to put bonds across the glide plane so the top and bottom are coupled. Then

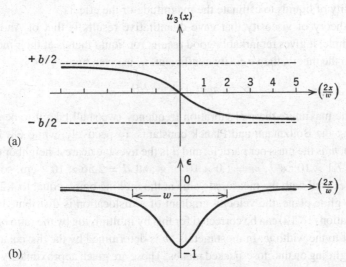

Figure 18.3 Distribution of displacement and shear strain along a glide plane near the center of a dislocation: (a) variation of displacement $u(x)$ from $-b/2$ to $+b/2$ for a stationary dislocation; (b) variation of shear strain, ϵ is the displacement gradient, $\epsilon = du/dx$, from 0 to -1 and back to 0. The width of the distribution is w. For a moving dislocation (velocity v), this figure indicates the velocity gradient $d\epsilon/dt = (v/a)(du/dx)$, equal to the strain rate along the glide plane at the core of a moving dislocation; a is the thickness of the glide shear region. Note that both sketches neglect perturbations caused by the local atomic structure.

the top is slowed by the bottom and vice versa. The amount of momentum transferred in this case is given by the drag stress times the time involved (which is b/v_d).

18.5.3 Simple metals

Very pure simple metals (those bound primarily by s- and p-level electrons) are very soft because of their delocalized bonding. Their flow stresses are essentially independent of temperature for the same reason. However, dislocations in them move through two quasi-fluids: the electron fluid and the phonon fluid. It is sometimes said that they experience drag from the electron and phonon "winds". These resistances are small, but measurable (Nadgornyi, 1988).

Although most discussions of these resistances are in terms of the interactions of the phonon and electron winds with the elastic strain fields of moving dislocations, it has been shown that the integrated effects are approximately eight times larger at the dislocation cores because the strain rates are so much larger there compared with the strain rates in the elastic fields (Gilman, 1968a).

In the case of phonons, as they cross the glide plane of a dislocation moving with a velocity v_d (and it is a positive dislocation so the material above the glide plane moves faster than that below it), on average they will carry more momentum down the velocity gradient, v_d/a (where a is the glide plane spacing), than up it. This is the same process that occurs in liquids in a somewhat less organized way. Therefore, we can use a simple theory of the viscosity of liquids to estimate the magnitude of the effect.

The first theory of viscosity that gave quantitative results is that of Andrade (1934). Although simple, it gives remarkably good results for liquid metals at their melting points. According to the theory, the viscosity coefficient, η, is given by:

$$\eta = 4/3(vm/d) \tag{18.12}$$

where v is the maximum thermal vibration frequency (we shall take it to be $k\Theta_D/h$ with k and h being the Boltzmann and Planck constants, respectively, while Θ_D is the Debye temperature), m is the mass per particle, and d is the average nearest-neighbor distance. For copper, $v = 7.1 \times 10^{12}\,\text{s}^{-1}$, $m = 1.0 \times 10^{-22}$ g and $d = 2.56 \times 10^{-8}$ cm, so $\eta = 4.7$ cP. This compares well with the measured value at the melting point, equal to 4.2 cP.

Along the glide plane, the velocity gradient of a dislocation is distributed as in Figure 18.3(b). Equation (18.12) can be corrected for this by multiplying by the ratio of the Burgers displacement to the width w. In most metals, w is determined by the dissociated widths of dislocations gliding on the close-packed planes. These are given approximately by (Cottrell, 1953):

$$w = [Gb^2/8\pi\gamma_{sf}][(2 - v)/(1 - v)] \tag{18.13}$$

where γ_{sf} is the stacking fault energy, and v is Poisson's ratio.

Applying this to copper, using 64 amu for the atomic mass of copper, a Debye temperature of 327 K, $d = b = 2.56$ Å, $G = 0.31$ Mbar, $\gamma_{sf} = 73$ erg/cm^2, and Poisson's ratio $v = 1/3$, the results are $w = 24$ Å, $\eta = 4.2 \times 10^{-3}$ P; this may be compared with the value measured by Greenman, Vreeland, and Wood (1967) of 7.2×10^{-4} P. Since all of the models involved are crude, the agreement is acceptable. There is also reasonable agreement with the values of $\eta = 1.4 \times 10^{-4}$ P deduced from internal friction data as given by Kobelev, Soifer, and Al'shits (1971).

At temperatures below the Debye temperature, the phonon concentration decreases until only the zero-point phonons are left at $T = 0$ K, and in the simple metals nearly free-electron liquids. The viscosity of the electron gas itself (due to electron–electron scattering) is small. The electron–ion scattering is also small, so electron drag is masked by the viscosity of the phonon gas at temperatures above the Debye temperature. However, in a nearly perfect simple metal at very low temperatures where the phonon–phonon scattering becomes negligible, the electron drag becomes dominant. Its existence is clearly demonstrated by the marked increase in dislocation mobility (in Nb and Pb) that is observed when the temperature passes below the superconducting transition temperature (Kojima and Suzuki, 1968). Further evidence is provided by the effects of magnetic fields on the flow stress of Cu (which is not a superconductor) at low temperatures (Galligan, Lin, and Pang, 1977). The magnetic effects appear to be caused by electron–dislocation scattering (or the lack of it), rather than the electron–electron viscosity originally proposed by Mason (1968). They are similar to magnetoresistance, except that they are strikingly anisotropic (Galligan *et al.*, 1986) because the cyclotron orbits of the electrons in the presence of the magnetic field are quantized.

A large amount of effort has gone into estimating electron–dislocation scattering coefficients. The work done up to a decade ago has been reviewed by Nadgornyi (1988). The important factors are easily identified. Only electrons at the Fermi surface can be scattered, and they have velocities v_f, so their momenta are $m v_f$. Since the Fermi wavelength is of the order of b (the Burgers displacement), the scattering cross-section is of order b (per unit length of dislocation) so the force generated in a scattering event is approximately $bm v_f$. The electron concentration is n, so the total drag on a moving line is proportional to $nbm v_f$ and the electron drag parameter (viscosity) $B_e = Cnbm v_f$ where C is a constant of order unity. For copper, $b = 3.6$ Å, $v_f = 1.6 \times 10^8$ cm/s, $n = 8.5 \times 10^{24}$ cm^{-3}, and $m = 9.1 \times 10^{-28}$ g, yielding $B_e \approx 3 \times 10^{-4}$ P, or less if $C = 1$ is too large. Thus, the simple Andrade theory again gives a straightforward interpretation of the observed damping. At low temperatures the effective viscosity rises (in the normal state) because the electrons carry momentum over longer distances than d.

18.5.4 Anisotropic metals

Most of the Periodic Table of the Elements consists of metals. The number of identified elements increases year-by-year so defining a total number is arbitrary. Let us take the first 100 of them. About 18 of these are non-metals (taking As, Se, Sb, and Te to be metals) so 82% are metals. The average number of valence electrons is about four, and these can occupy a large number of both pure and hybrid states near the Fermi surface of a given element. Thus the number of possibilities is truly staggering. For structural applications, since electron density is so important for strength, the strongest materials will have symmetries no greater

Table 18.1 *Stacking fault energies at room temperature*

Element	γ_{sf} (erg/cm^2)
Aluminum	166
Cadmium	175
Cobalt	15
Copper	78
Gold	45
Iridium	300
Magnesium	125
Nickel	128
Palladium	175
Platinum	322
Silver	22
Tungsten	>50
Zinc	140
Zirconium	240

than cubic, although cheaper materials such as glasses may be nearly isotropic. Thus most of the materials of interest will be anisotropic.

Anisotropy takes several forms.

(1) *Glide plane* – in a close-packed atomic structure, the bonding energies of the second-nearest neighbors may be significant. This may result either from proximity, or from symmetry, effects. The effect may be either positive (bonding) or negative (anti-bonding). It may make the energy difference between f.c.c. (stacking sequence ababab) and h.c.p. (hexagonal close-packed, stacking sequence abcabc) very small. Then unit dislocations will have a strong tendency to become extended along their glide planes providing a preference to the set of planes on which plastic flow first occurs. A prototype for this is β-brass crystals. Table 18.1 lists a few values of stacking fault energies according to Murr (1975).

Another anisotropy of glide planes is the bonding across them. For example, crystals with hexagonal symmetry are usually not quite close packed. Therefore, the chemical bonds perpendicular to the close-packed planes are not the same as those parallel to these planes. As a result, for example, zinc crystals (Gilman, 1956) glide much more easily on (0001) close-packed planes than on prismatic planes such as (01$\bar{1}$0). Similar results were obtained for cadmium (Gilman, 1961). At a given temperature and strain rate, the difference in the stresses for glide on the two planes in Zn and Cd is a factor of about 1000. This is far too large to be accounted for by the Peierls theory, or some other theory involving pair potentials. A chemical theory is necessary.

(2) *Glide direction* – since the energy of a dislocation is proportional to the square of the magnitude of its Burgers vector, the glide direction is essentially always the shortest direction of translation symmetry in a given crystal structure. Sometimes this is ambiguous as it is near order–disorder transitions (compositions or temperatures), but these cases are rare.

(3) *Sense of glide direction* – a dislocation's elastic energy is independent of the sense of its Burgers vector, but its mobility need not be because its core may not have mirror symmetry

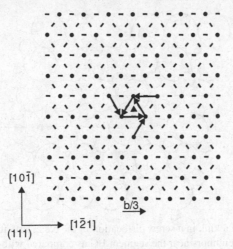

Figure 18.4 Possible three-fold splitting of the core of a screw dislocation in the b.c.c. crystal structure.

about its mid-point. It may have "saw-tooth" symmetry, so it behaves like a ratchet, that is, unidirectionally. One form of this asymmetry appears in b.c.c. metals where the glide direction is ⟨111⟩ which has three-fold symmetry (as contrasted with two- or four-fold). Therefore, instead of extending on a single glide plane, screw dislocations parallel to ⟨111⟩ with (1/2) ⟨111⟩ Burgers vectors can extend on three planes equally (Sleezwyk, 1963). Each of these is a vector of the form $(a/6)[111]$, and they lie on the three planes: $(11\bar{2})$, $(12\bar{1})$, and $(21\bar{1})$. Figure 18.4, after Vitek (1974), is a schematic representation of the atomic configuration. It may be seen that shearing this configuration from left to right is different than shearing it from right to left. While the details of a calculation such as that of Figure 18.4 are not reliable, since the potentials used are quite approximate as applied to dislocation cores, the general idea that three-fold splitting occurs is important.

According to calculations by Suzuki (1968) the resistance to dislocation motion caused by the crystal structure of a b.c.c. metal is small, especially when the zero-point energy is taken into account. As a consequence, the dislocation lines can be expected to meander through the structure, wandering from one direction to another on given planes, and from one plane to another. This is consistent with the glide offset markings that are observed on the surfaces of deformed b.c.c. crystals.

18.5.5 Transition metals

In the transition metals the electrons are no longer nearly free. Those that contribute most to the cohesion are localized in spd-hybrid bonds. Therefore, the cohesive energy is not nearly independent of the atomic configurations, and the simple models discussed above do not apply. The behavior becomes complex, and theoretical descriptions of it tend to become ad hoc. No attempt will be made here to describe a theory for these metals, although they are the most important ones for practical uses.

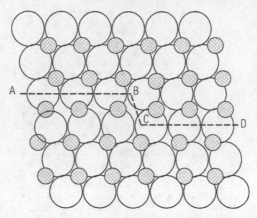

Figure 18.5 Plan view of a kink in a screw dislocation in the b.c.c. structure. Note the change in the positions of the nearest neighbors near the segment BC as compared with AB and CD.

It is commonly believed that dislocation mobility becomes small in the b.c.c. transition metals at low temperatures. However, observations of rising yield stresses at low temperatures are probably a result of small concentrations of interstitial impurities. As these metals are made increasingly pure, their yield stresses at low temperatures become increasingly small. Also, it is very difficult to remove the last traces of impurities in them. Some of the purest specimens were made by Powers (1954), and they were exceptionally soft at low temperatures.

Figure 18.5 is a view looking downward onto the (110) glide plane of the b.c.c. structure. A ⟨111⟩ glide direction lies horizontally in the figure. Two planes, one above (shaded circles) and the other below the plane of the paper (open circles), are shown. The top plane has partially glided over the bottom one creating a dislocation line, ABCD, that is moving from the top to the bottom of the diagram. Furthermore, the dislocation line has an offset at approximately its mid-point. The offset (kink) has deliberately been taken to be sharp because such a configuration will present a maximum resistive force towards movement of the kink. More gradual kink configurations will present smaller resistive forces. This diagram is highly schematic, but it makes the point that the atomic configuration at a kink is very different than it is in the normal crystal structure. Note also that there is little change in the atomic density at the kink. This suggests that the valence electron density may not be significantly affected, so the dislocation's energy may not vary much with the position of the core. Therefore, the mobility may be high, consistent with the experimental observations.

The extensive cross-gliding that occurs during the deformation of b.c.c. metals provides further evidence that dislocation energies in them are nearly independent of position.

Since the b.c.c. structure is not close packed, it might be expected that b.c.c. metals would have directional bonds which would lead to low dislocation mobilities at low temperatures. However, Frank (1992), as mentioned in Chapter 13, has pointed out that the extra volume in the b.c.c. structure (relative to the f.c.c. structure) tends to lower the kinetic energies of the valence electrons, and it is this effect that stabilizes the structure rather than directional bonds.

Direct evidence of high dislocation mobilities in pure b.c.c. metals has been provided by the observation that the yield stresses of these metals are affected by applied magnetic fields at low temperatures (Galligan *et al.*, 1986). This means that their yield stresses are affected by the scattering of nearly free electrons which strongly suggests that their motion through the matrix is underdamped part of the time.

Less direct evidence of high mobilities in high purity b.c.c. metals are the observations of Powers (1954).

In contrast to ideal metals, commercial transition metals do resist dislocation motion. Average mobilities in them are low, sometimes very low. This results from extrinsic effects, including impurities (both interstitial and substitutional), other dislocations (as well as dipoles), precipitates (and Guinier–Preston precursors), grain boundaries, stacking faults (and anti-phase boundaries), short-range order, short-range disorder, vacancies and self-interstitials, domain walls (both electric and magnetic), and others.

18.5.6 Ionic compounds

Very high purity ionic crystals such as the alkali halides are very soft. That is, dislocations move through them with little restraint. For example, pure NaCl crystals are so soft that the small thermal stresses produced by evaporative cooling after they have been dipped in alcohol are enough to cause substantial amounts of dislocation motion in them. Also, high purity LiF crystals cannot be cleaved, while less pure ones cleave readily. Since ionic crystals do not contain free electrons, the only intrinsic drag mechanism in them is that caused by the phonon wind.

The many extrinsic sources of drag that are present in metals also act in ionic crystals. In addition, there are some drag sources associated with local charge repulsion. Thus local charge neutrality is important. The shells of neighboring ions (first nearest, second nearest, third nearest, etc.) surrounding a given ion alternate in the signs of their charges in ionic crystals. To maintain this arrangement during plastic shearing, only selected combinations of shear directions (and planes) are allowed. Otherwise, glide (shear) brings ions of the same sign into nearest-neighbor positions. Adjacent charges of the same sign have very high energies. Therefore, high dislocation mobility is intrinsically limited to the favorable directions and/or planes. Gliding motions in other glide systems (plane plus direction) require much higher applied stresses, unless the temperature is high where increased polarizability tends to screen the local charges.

The most common crystal structure for ionic compounds is the rocksalt, or NaCl, structure. Figure 18.6 shows the glide plane and glide direction in this structure. The glide plane is (110) and the shear direction is [1$\bar{1}$0]. For this combination, it can be seen that the shear on the glide plane occurs between ions of opposite sign in a direction in which all of the ions have the same sign. Also, for compression (or tension) applied along a ⟨100⟩ direction, the illustrated glide plane would lie at an angle of 45° with respect to the principal stress

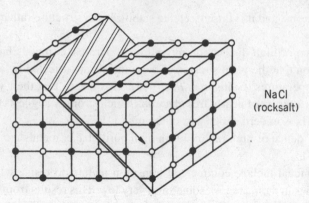

NaCl
(rocksalt)

Figure 18.6 Schematic illustration of the primary glide plane (101) and the primary glide direction [10$\bar{1}$] in the rocksalt (NaCl) crystal structure. Note that the glide direction lies parallel to rows of ions of the same charge sign.

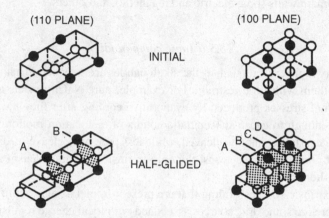

(110 PLANE)   (100 PLANE)

INITIAL

HALF-GLIDED

Figure 18.7 Illustrating the effect of the glide plane (at constant glide direction) on the configuration of the ions at the center of an edge dislocation core in the rocksalt crystal structure. The initial and final configurations with $u(x) = 0, b$ are shown at the top, while half-glided configurations with $u(x) = b/2$ are shown at the bottom. Note that the nearest-neighbor ions of the planes A and B have opposite charge signs in the case of the (110) glide plane, whereas the nearest-neighbor ions of the A, B, C, and D planes have the same charge signs in the case of the (100) glide plane. In both cases the glide direction is parallel to rows of ions having the same signs.

axis, making it the plane of maximum shear stress, and making the overall configuration the "easy glide system". This is the system of highest mobility for rocksalt-type crystal structures (mainly the alkali halides). It is the primary glide system.

If ionic crystals are very pure, the mobilities of dislocations in their primary glide systems are very high. Thus there is little intrinsic resistance to their motion. The situation is quite different, however, for motion on other glide systems. Suppose that the principal stress axis for an ionic crystal is changed to ⟨111⟩, either by cutting a specimen of that orientation, or by applying torsion about one of the [100] axes. Then the shear stress on the {110} planes is zero, and the shear stress becomes maximized on the {100} planes. However, in order

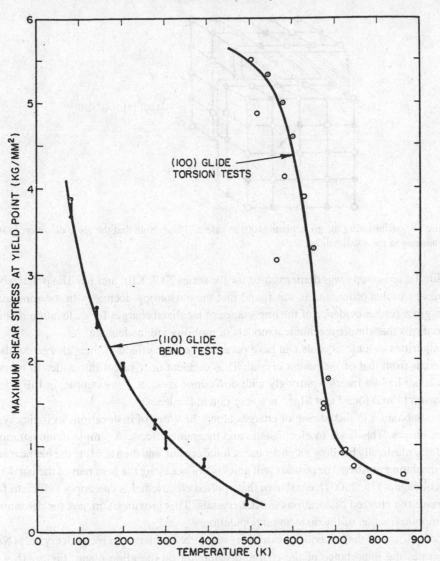

Figure 18.8 Plastic anisotropy of lithium fluoride. Note that at 500 K the yield stress is five times larger for glide on the (100) planes of LiF compared with the (110) planes. In both cases, the Burgers vector is along a face-diagonal.

for plastic shear to occur on the (001) plane (still in the [1$\bar{1}$0] direction) the configuration at the core of the dislocation causes ions of the same charge sign to become nearest neighbors (Figure 18.7). This markedly reduces the mobility, raising the stress needed for a given velocity from nil to a large value. This plastic anisotropy has been measured for LiF crystals (Gilman, 1959). Some results are shown in Figure 18.8. Also, a theory of the indentation hardnesses of alkali halide crystals, based on this phenomenon, has been developed (Gilman, 1973).

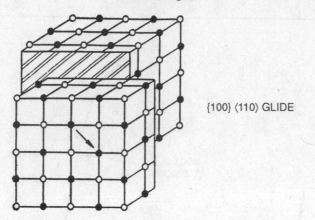

{100} ⟨110⟩ GLIDE

Figure 18.9 Illustrating the glide plane (100) in galena (PbS). Note that the glide direction {110} is the same as in the alkali halides.

Similar anisotropy was demonstrated for the series KCl, KBr, and KI. The polarizability increases in that order, and it was found that the anisotropy decreases in the same order. This gives further evidence of the importance of localized charges for dislocation mobility. It confirms that simple mechanical models of mobility are inadequate.

Impurities in ionic crystals can have particularly large effects if their charge number is different from that of the parent crystal. Thus divalent or trivalent impurities in monovalent alkali halides interfere strongly with dislocation motion. For example, in LiF crystals Johnston (1962) found that Mg^{2+} is a very potent hardener.

An imbalance in the pairing of charges along the cores of dislocations in ionic crystals often occurs. This leads to electrostatic and magnetic effects. A simple demonstration is made by plastically bending a slender bar of ionic crystal, and then holding the bar near some dry insulating powder. The powder will quickly collect along the bent part of the bar. Li and his colleagues (Li, 2000) have shown that applied electric fields can apply sufficient force to move the charged dislocations in KCl crystals. This provides a means for determining the mobile fraction of the dislocation population.

Comparison of the behavior of galena crystals (PbS) with that of rocksalt crystals (NaCl) illustrates the importance of the charge distribution on the glide plane. Figure 18.9 is a schematic drawing of glide in galena. The crystal structure, and the Burgers vector, are the same as for rocksalt. However, the bonding is delocalized since this compound has metallic conduction. Therefore, the glide plane is not restricted to one containing only rows of ions of the same sign, and the dislocation cores can lower their energies by gliding between the most widely separated planes in the structure. ⟨110⟩ remains the primary glide direction, but the primary glide plane becomes {001} because the charge conflict does not occur.

18.5.7 Carbides (and other "hard" metals)

Transition metal carbides, nitrides, and borides are very hard compared with the metals from which they derive, although they are metallic conductors of electricity. In part this is

Table 18.2 *Yield stresses of carbides compared with metals*

				Yield stress (GPa)	
Pure material	B (GPa)	G (GPa)	b	Observed	Peierls theory
SiC	224	136	3.07	12.0	0.019
TiC	242	187	3.06	11.0	0.026
Ni	180	63	2.49	0.0033	0.009
Ta	196	60	2.06	(0.0004)	0.008

because the concentration of valence electrons in them is unusually high. This is a result of the metalloids (C, N, and B) "fitting" into the interstices of the transition metal arrays, thereby increasing the number of valence electrons and quantum states with only moderate increases in the volumes of the unit cells. Various hybrid covalent bonds can then form. This disproportionately increases the shear moduli relative to the bulk moduli as Figure 13.9 shows, a necessity for high hardness. The average G/B ratio for the six metals in the figure is 0.37, while it is 0.71 for the six carbides. Thus the carbides are twice as "rigid" as the corresponding metals, although their ratios are not as high as that of diamond (maximum $G/B = 1.21$). The overall stabilities of the carbides are large as indicated by their bulk moduli. This inhibits the formation of dislocation cores, while the large G/B ratios suggest that the core energies depend strongly on their positions.

Why are crystals such as SiC and TiC very hard, while metals with comparable overall cohesive energies, and purities, are orders of magnitude softer? Mechanical models like the Peierls theory do not give a satisfactory answer, either qualitatively or quantitatively. As in the case of covalent crystals, approaches used in chemistry to deal with complex reactions seem best.

Table 18.2 summarizes the problem. It compares the essential mechanical properties of the two prototype substances, SiC and TiC, with two metals, Ni and Ta, of similar cohesive energies. The four bulk moduli, B, are roughly the same. The shear moduli (on the glide planes) G vary more, being distinctly smaller for the metals, but all are of the same order of magnitude. The magnitudes of the Burgers vectors, b, are roughly the same for all four. However, the yield stresses are very different, by 4–5 orders of magnitude! Also, the values given by the simple version of the Peierls–Nabarro theory are 3–4 orders of magnitude off the mark.

Strong evidence that chemical (electronic) factors are important is provided by the correlations that exist between the cohesive properties and the electronic band structure.

It was proposed by Cohen (1991) that since hardness and bulk moduli tend to be proportional, the latter is a guide to finding harder materials. Table 18.2 indicates that this guide is not reliable. Note that SiC is harder than TiC, but the latter has the larger bulk modulus. Furthermore, the reasoning behind this idea is faulty because it is the shear modulus that measures the resistance of a solid to shape changes. The bulk modulus measures resistance to volume changes. The latter have little direct connection with hardness, as is apparent

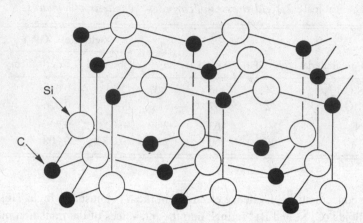

Figure 18.10 Elevation view of the structure of cubic SiC. The two "planes" in the drawing represent the (111) planes of the structure. Gliding occurs between two such planes.

from the fact that it is the shear modulus rather than the bulk modulus that appears in the equations of dislocation theory. There is an empirical proportionality between the bulk and shear moduli, but it is far from reliable, as Table 18.2 indicates. A better empirical correlation between indentation hardness values and shear moduli has been demonstrated by Teter (1998).

18.5.7.1 Silicon carbide

The strong temperature dependence of the plasticity of SiC indicates that the rate-determining process is localized to atomic dimensions, that is, to kinks along dislocation lines. The motion of a kink in SiC is best viewed from above, normal to the {111} glide plane. The variant of SiC that will be discussed is the cubic one, but similar considerations should apply to the hexagonal variant. Cubic SiC has the zinc blende structure, and an elevation of this is presented in Figure 18.10. The plan view of the structure with a kinked dislocation line is shown in Figure 18.11. Notice that the structure consists of alternating layers of Si and C atoms where the patterns in each layer consist of sets of triangles. Also note that in the structure a plane of carbon atoms faces a plane of silicon atoms across the {111} glide plane. It seems likely that gliding in this structure occurs where the density of chemical bonds is smallest, and the separation of the atoms is largest. Figure 18.11 shows a region straddling a (111) glide plane where the upper plane of atoms has partially slid over the lower plane thereby creating a dislocation that lies on the (111) plane. This line contains a sharp kink, BC. The shaded circles represent atoms (carbon) that lie just above the plane of the diagram, and the open circles represent the plane of atoms (silicon) just below the plane of the diagram. The line, ABCD, represents a screw dislocation line. When it moves downwards one atomic spacing, the top layer of atoms moves sideways to the right by one spacing.

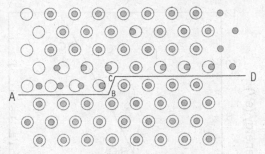

Figure 18.11 Plan view of the glide plane of the SiC structure with a screw dislocation AD and a kink BC. The open circles represent atoms just below the glide plane. The shaded circles represent atoms just above the glide plane. The upper plane of atoms above the line AD has moved one spacing to the right with the center of action being the kink at BC.

The top layer does not move concertedly because it is easier for it to move with the aid of a kink, or series of kinks. If the kink concentration (fraction of possible atomic sites) is c_k, and each kink has the same velocity v_k, then the velocity of the dislocation line is $v_d = c_k v_k$. The kink starts at a position where the SiC bond is intact, moves one atom step, and finishes at a position where another bond is intact (having exchanged one atom for another). Figure 18.11 shows the kink, BC, at its mid-glide position where the local chemical structure is maximally disturbed.

Motion of the kink by one atomic distance involves separating one Si–C pair at the front of the moving kink, and then reforming a pair using different atoms behind the kink. Thus, the process is analogous to a chemical substitution reaction:

$$SiC + C' \rightarrow (CSiC')^* \rightarrow SiC' + C$$

where $(CSiC')^*$ is the activated reaction complex. This reaction is embedded in the open zinc blende structure, but is very localized. This is expected because of the highly directional hybrid bonds.

Electronic band structure is also important for the motion of kinks in SiC. This is consistent with Figure 18.12 which relates activation energies for glide with the gaps in the bonding energy spectrum (HOMO–LUMO, or band gap). The point for SiC correlates very well with the data for the Group IV elements. The figure indicates that its glide activation energy is twice its LUMO–HOMO gap.

Other carbides, such as TiC and WC, exhibit metallic electrical conduction, so their bonding is not entirely covalent. It follows that the barrier to kink motion in these crystals is determined by the overall stabilities of these crystals (measured by their free energies of formation and their vibrational energy spectra).

The fact that the dislocation velocity is approximately linearly proportional to the applied stress indicates that there is efficient momentum transfer across the glide planes during the motion.

The energetics of this type of reaction will be discussed in more detail in connection with the behavior of the homopolar crystals, Si and C.

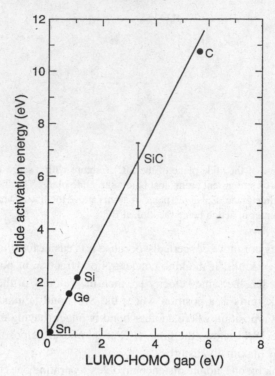

Figure 18.12 Correlation between glide activation energies and LUMO–HOMO energy gaps (i.e., band gaps) for Group IV crystals, showing that the former equals twice the latter.

At high temperatures, the barrier to dislocation motion in covalent crystals is the activation energy. At low temperatures, two cases arise. In tension, the mobility is expected to decrease with decreasing temperature, and the flow stress to continue to rise until fracture intervenes. In compression (including indentation hardness), shear-induced transformation to the β-tin or some other structure is encountered (Gilman, 1993a), and the indentation hardness becomes nearly independent of temperature.

Since SiC is not centrosymmetric there is a piezoelectric effect in it which may play a secondary role by altering the energy levels locally through the creation of local electric fields. This will also tend to cause a polarity effect (difference in mobility of dislocations with half-planes ending on anions versus those with half-planes ending on cations).

18.5.7.2 Titanium carbide

Because of its metallic conductivity, there is no LUMO–HOMO gap for TiC although its bonding is dominated by covalent bonds (pd-hybrid orbitals) between its carbon and titanium atoms (Price, Cooper, and Wills, 1992). The glide plane in TiC is shown schematically in Figure 18.13. The structure is the rocksalt structure and the figure shows it with a (111) glide plane exposed, including a row of three carbon atoms lying just above the glide plane

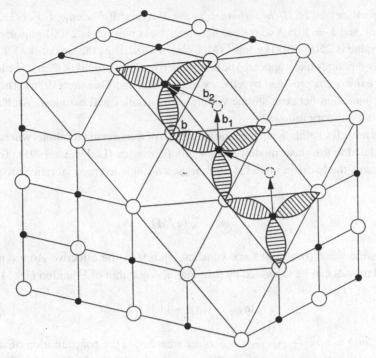

Figure 18.13 Schematic elevation of the TiC structure (rocksalt) with a (011) glide plane exposed. A row of three carbon atoms lying just above the glide plane (solid circles) is shown. The lozenge-shaped objects represent the C–Ti covalent bonds. The motion of a glide kink carries a carbon atom from one position to another indicated by the Burgers vector **b**. For either the direct path, **b**, or the indirect path, $\mathbf{b}_1 + \mathbf{b}_2$, when the carbon atom is in the mid-glide position, $\mathbf{b}/2$ or $\mathbf{b}_1$, the symmetry of the covalent bonds is broken, so there is a chemical activation barrier. The most likely mid-glide position is $\mathbf{b}_1$ where the coordination number of the carbon atom is preserved, but instead of octahedral symmetry it has trigonal prismatic symmetry.

(solid circles). The lozenge-shaped objects represent the C–Ti hybrid covalent bonds. The motion of a glide kink carries a carbon atom from one position to another indicated by the Burgers vector, **b**. For either the direct path **b**, or the indirect path $\mathbf{b}_1 + \mathbf{b}_2$, when the carbon atom is in the mid-glide position, $\mathbf{b}/2$, or $\mathbf{b}_1$, the symmetry of the covalent bonds is broken, so there is a chemical activation barrier of about 2.2 eV. The symmetry changes from approximate octahedral symmetry to approximate trigonal prismatic symmetry, and then returns to octahedral symmetry after a unit of motion has occurred.

From the accepted electronic structure (Price and Cooper, 1989), the difference in energy between the average of the anti-bonding states and the Fermi energy is about 2.6 eV. This is comparable with the chemical heat of formation (44 kcal/mol = 2.2 eV).

The bonds are strongly disturbed by the motion of kinks through the structure. It has been shown elsewhere that this leads to a simple theory of the hardness in terms of the enthalpy of formation, ΔH_f, of the bonds (Gilman, 1970). The expression for the hardness (low temperature) is $2\Delta H_f/b^3$, where b is the Burgers displacement. For TiC, ΔH_f is 44 kcal/mol, and $b = 3.05$ Å so the predicted hardness number is 2500 kg/mm^2, while the observed value is 2500–3400 kg/mm^2 (McColm, 1990). Thus the hardness of this carbide and other compounds like it appear to be determined by the strengths of their chemical bonds.

Further evidence is provided by a theory of Grimvall and Thiessen (1986). They demonstrated a connection between atomic vibration frequencies and hardness. An abbreviated version of their theory follows.

Specific heats for solids depend on the excitation of transverse vibrations whose frequencies are related to the shear moduli and the atomic masses (Ledbetter, 1991). To simplify the discussion, the Einstein model is used. It uses a single average vibration frequency ω_e given by:

$$\omega_e = \sqrt{(g/M)} \tag{18.14}$$

where g is the shear (bending) force constant, and M is the effective atomic mass. The forces and masses can be separated by forming the logarithm of Equation (18.14):

$$\ln \omega_e = (\ln g - \ln M)/2 \tag{18.15}$$

$M = \sum(c_j m_j) = c_1 m_1 + c_2 m_2 + \cdots + c_n m_n$ where c_j is the concentration of substituent j, m_j is the atomic mass of substituent j, and n is the index of substituent. The thermal energy of an Einstein oscillator is $k\theta_e$ where k is Boltzmann's constant and θ_e is the Einstein temperature. The mechanical energy of the oscillator is $\hbar\omega_e$ where $\hbar = h/2\pi$ with h Planck's constant. Then an entropic characteristic temperature θ_s can be defined by the relation:

$$\ln(k\theta_s) = \langle \ln(\hbar\omega_e) \rangle \tag{18.16}$$

and an effective force constant, g using Equation (18.14):

$$k\theta_s = \hbar\sqrt{(g/M)} \tag{18.17}$$

By analyzing entropy versus temperature data, values for θ_s can be obtained, and from them values of g. Just as the bulk moduli depend on valence electron densities, so do the values of g (Grimvall and Thiessen, 1986). The dimensions of g are dyn/cm. Therefore, to obtain characteristic vibrational energy densities, the g values need to be divided by the lattice parameter, a. Then, g/a has the units of energy per unit volume, or GPa, which is the same as the units of indentation hardness, H. The correlation between H and (g/a) is shown in Figure 18.14 for some carbides and nitrides. This figure is similar to Figure 3 of Grimvall and Thiessen (1986). This plot is better than plotting H versus B (bulk modulus) because the fact that a compound contains atoms of more than one mass has been taken into account.

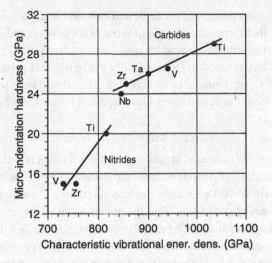

Figure 18.14 Plot of hardnesses of carbides and nitrides versus force constant parameters (units of modulus) derived from entropic specific heat (after Grimvall). These are averaged force constants, but they are characteristic of particular materials. All of the crystals here have the rocksalt structure. The hardness data are from Teter (1998).

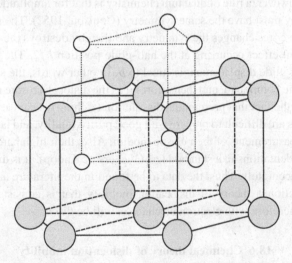

Figure 18.15 Crystal structure of WC consisting of hexagonal planes of W atoms (shaded circles) stacked directly above one another with carbon atoms occupying one-half of the interstices in the trigonal prisms of W atoms. In this case, the center of a kink has octahedral symmetry. Thus, the TiC and WC structures are conjugates.

18.5.7.3 Tungsten carbide

The crystal structure of WC is based on trigonal prisms (Figure 18.15). It consists of close-packed planes of W atoms lying directly over one another to form hexagonal cells containing eight atoms at the corners (one atom per cell), and two interstitial holes at the centers of the two trigonal prisms that form the cell. One of these holes is occupied by a carbon atom.

When a kink in this structure lies at its mid-glide position, the symmetry of the hole becomes approximately octahedral (from initially being trigonal prismatic). Thus the dislocation cores in WC are conjugates of those in TiC. Resistance to the motion of the dislocations comes from the destruction of the chemical bonding within the cores.

18.5.7.4 Other "hard" metals

These are combinations of transition metals and metalloids (typically B, C, and N). They are also known as interstitial compounds because the metalloid atoms fit into the interstices of the transition metal frameworks. Usually, the interstitial holes have the symmetry of either octahedra or simple hexagonal prisms. Thus the transition metal frameworks are either f.c.c. arrays or simple hexagonal arrays. The resulting compounds are 2–4 times harder than the corresponding commercial metals. These very hard compounds conduct electricity about as well as resistive alloys (e.g., Nichrome), and sometimes as well as pure metals. They are used to make cutting tools and wear-resistant surfaces.

Evidently, the low mobilities of dislocations in hard metals has a chemical basis as indicated above. The compounds have high heats of formation, so their crystal structures are very stable; however, a rule of quantum chemistry is that for amplitude (wave) functions to form bonds they must have the same symmetry (Coulson, 1952). The shear deformation within dislocation cores changes the symmetry and thereby destroys the chemical bonding with the maximum effect occurring at the half-glide position $b/2$. The initial structure is restored when the glide displacement is equal to b. In other words, the symmetry changes first decompose the compound, and then reform it as the dislocation core moves along. This is consistent with the fact that preferential dissolution (etching) occurs at dislocation cores.

The hard metals are difficult to prepare with good purity, quality, and large enough size to make accurate measurements of their elastic moduli. Also, their high hardnesses yield very small hardness indentations in a standard test (Vickers or Knoop). It is difficult to measure the size of these accurately. Thus, the data to be found in the literature are quite variable.

In the next section a theory of dislocation mobility that is consistent with quantum mechanics, and therefore with electron mechanics, will be discussed.

18.6 Chemical theory of dislocation mobility

Covalently bonded solids differ from metals and ionic solids in a qualitative way. They are hard instead of soft even when their purity is high. In other words they possess intrinsic plastic resistance. Most mineral substances are intrinsically hard at low temperatures. A few are soft such as native copper and gold. Also, some such as talc, asbestos, and mica are partially soft depending on the orientation of a stress tensor relative to their crystal structure axes. Minerals are mostly compounds, rather than pure elements, and when they are decomposed, soft metals are freed from them. The hardest compounds tend to be those with the highest melting points and heats of formation. These facts suggest that hardness and strong chemical bonds are related. It follows, since the strongest chemical bonds are those in which electrons with opposite spins are paired, and spin-pairing is a quantum mechanical effect, that hardness is explicitly a quantum mechanical effect. Hardness cannot be understood in terms of classical mechanics.

Several prototype minerals remain hard as the temperature is raised until a critical temperature is reached; then they quickly soften with further increases in the temperature. Classical thermodynamics does not explain this behavior. Quantum statistical mechanics is required.

All of these behavioral factors can be understood if kink motion is considered in terms of local (embedded) chemical reactions, as contrasted with mechanical processes (Gilman, 1993b).

A dislocation line marks the boundary between two areal regions of a glide plane. In one region sliding (translation) of the crystal on one side (say the top) of the plane has occurred relative to the part of the crystal that lies on the other side (the bottom). In the other region, sliding has not occurred. In crystals, the amount of sliding is quantized in terms of the translation vectors of the structure, so the registry of the crystal is restored after the sliding. However, there remains a ribbon (or zone) of disregistry locally at the central position of the dislocation line (the core). The motion of dislocations is limited by the motion of their cores, and the core motion is limited by the motion of kinks along the cores. Thus, a dislocation line can move no faster than its kinks.

18.6.1 Group V elements

The core chemistry at kinks needs to be considered in some detail. In materials with localized bonding, dislocations are expected to move disconcertedly, that is, bond by bond. At an abrupt kink, the chemical structure of a crystal is severely disrupted (a chemical bond is "broken"). The disruption is very localized, to a volume of the order of one atomic (or molecular) volume in size. The elastic fields of these kinks are similar to those of "point defects" according to Saint-Venant's Principle. Their elastic strength declines very rapidly with distance, r, in proportion to $1/r^3$. Thus at distances of about twice the Burgers displacement, b, kinks interact only weakly since the strength is about eight times less strong.

When a kink moves, it breaks the atomic bonding symmetry, a process forbidden by the rules of quantum mechanics unless there is a large energy change. Numerical calculations using fixed interatomic potentials are not reliable because the symmetry changes cause changes in the local potentials. Also, the potential energies within a kink must differ from those outside it because the amplitude (wave) functions cannot remain continuous as they pass through a kink.

One of the difficulties of the classical Peierls–Nabarro theory is that the potential is periodic, and therefore continuous, but in real covalent crystals there is a singularity at each core, and particularly at each kink. Other difficulties are discussed by Nabarro (1997), and by Hirth and Lothe (1968). Its general deficiency is that a classical theory should not be expected to be applicable at the atomic scale.

The motion of a kink is analogous with a chemical substitution reaction, described by Woodward and Hoffmann (1970) as a disconcerted process (i.e., a forbidden one). If the average kink velocity along a dislocation line is v_k, and the kink concentration is c_k, then the velocity of the line v_d is $\alpha c_k v_k$, where α is a geometric factor of order unity. The conventional theory is based on the idea that the rate of formation of kink pairs is the rate-determining step, and that kink mobilities are always high (Alexander and Haasen, 1968). However, the usual theory of kink formation is questionable because dislocation cores are very localized, and inelastic, whereas the elastic field of a dislocation is very delocalized (it decays as $1/r$, as compared with $1/r^3$ for a kink). But the usual theory postulates a line energy of the core in terms of the energy of the elastic field. This is clearly incorrect (Gilman, 1993b). Direct measurements of kink formation rates have not been possible.

Perhaps the most convincing evidence of the failure of the conventional theory is that the behavior of crystals is not consistent with it. According to kink nucleation theory, if a crystal were flowing plastically at a high temperature, and the temperature were quickly decreased, there should be a transient period during which the crystal would continue to flow. This is not observed, according to Galligan *et al.* (1986).

The activation energies that are observed for dislocation motion are sharply defined, and independent of the applied stress (Sumino, 1989). This implies that the activation process is localized to atomic dimensions. If kink-pair formation were controlling the rate, according to the theory, the nucleation rate should depend on the applied stress, and there should be a spectrum of activation energies.

At low stress levels, the activation length for kink-pair formation is multiatomic. Since the temperature is typically well above the Debye temperature, the phonon wavelengths are short, and their phases are random. Therefore, the probability of having enough phonon correlation to reach the activated transition state is very small.

For the reasons just cited, it may be concluded that kink mobility, rather than formation rate, must control the overall rate.

In the open diamond structure, kinks can be very localized, consistent with the highly directional bonds, and the fact that the bonding electrons are paired in the molecular orbitals. Also, the fact that the dislocation velocity is approximately linearly proportional to the applied stress means that there is efficient momentum transfer across the glide planes during the motion. This yields substantial glide plane viscosity.

The activation energy for dislocation motion (Figure 18.12) equals twice the HOMO–LUMO energy gap (Gilman, 1975) for homopolar covalent crystals (HOMO highest occupied molecular orbital; LUMO lowest unoccupied molecular orbital). For homopolar crystals (C, SiC, Si, Ge, and Sn), the HOMO level equals the top of the valence band, and the LUMO level equals the bottom of the conduction band. This suggests that the kink mobility is directly related to the electronic structure. This is consistent with the idea that chemical factors control the mobility.

Other evidence that electronic structure is important in the flow of covalent crystals is provided by observations of unpaired electrons (free radicals) in plastically deformed Si crystals by means of spin resonance (Kveder *et al.*, 1991).

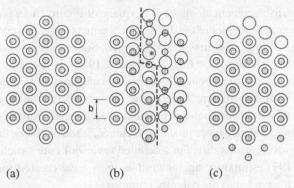

(a)   (b)   (c)

Figure 18.16 Diamond structure, screw dislocation with a kink. Open circles are below the glide plane, and shaded circles are above the glide plane. (a) Configuration before glide. (b) Glide proceeding from right to left, the dashed line indicates the screw dislocation of the Burgers vector **b**, and the asterisk marks the center of the kink. (c) Glide process complete; shaded circles have moved one spacing downward.

Consider the pertinent local kink geometry. Figure 18.16 is a sketch, looking from above, of a kink-pair on a screw dislocation line lying on a (111) plane of the diamond structure. Note the similarity to Figure 18.11. More detail of the chemical structures of kinks is shown by looking at a kink from the side (perpendicular to the plane of Figure 18.11), see Gilman (1993b).

The most disturbed part of the kink is indicated by an asterisk in the figure. Here, the immediate surroundings of the ions are changed markedly, including their coordination numbers. Compare this part near the asterisk with the structure far from the center of the kink; focus on the open circle at the asterisk. In the plane above it there are four nearest shaded circles. Away from the position with the asterisk, each open circle has one shaded circle directly above it, and six more shaded circles nearby in the upper plane. Thus the atomic structure at the center of the kink is very different from the normal crystal structure.

Another important feature is that the geometry of the crystal surrounding the kink is multiply connected, so geometric circuits taken around the dislocation line (normal to the plane of the drawing) undergo a phase shift of b each time a circuit of 2π is completed. This is expected to have a strong effect on the angular momenta of the bonding electrons by disturbing the continuities of the electron's probability amplitude functions.

The geometric complexities at kinks create great difficulties for the development of accurate quantum mechanical theories. Similar difficulties arise in the theory of free polyatomic molecules of substantial size. Therefore, arguments equivalent to those developed by quantum chemists for dealing with large molecules can be applied here.

The bonds across the (111) planes are sp^3-hybrid orbitals, especially for the diamond case. There are three other sp^3 bonds surrounding each atom. These bonds are quite resistant to shear distortions (bending) as discussed in Chapter 13. If they were not resistant to shear, the structure would collapse to some close-packed configuration.

The direction of displacement along the (111) plane of Figure 18.16 is a {110} direction. The maximum shear at the center of the dislocation's core is in what is commonly called the "shuffle" region, as contrasted with the "glide" region. Some authors consider that the maximum shear is in the "glide" region (e.g., Alexander, 1979). However, this seems unlikely since three times fewer chemical bonds lie normal to a unit area of the shuffle region, so this is the weaker region of the structure. Clearly, the valence electron density is smallest in the "shuffle" region, and it has been shown in previous chapters that strength is a monotonically increasing function of valence electron density. Hence, it seems likely that "shuffle" kinks are the ones that move most readily. For a detailed review of core structures see Duesbery and Richardson (1991). Impurity agglomerations may have created misleading electron micrographs which have led to the contrary opinions.

The essence of the present model is that the symmetry of the molecular orbitals plays a role that is similar to its role in determining the rates of chemical reactions according to the Woodward–Hoffmann theory (1970). Movement of a kink is then akin to a chemical reaction in which an embedded "molecule" is dissociated, and then one of the product atoms joins with an atom from another dissociation to form a new "molecule". That is, the reaction is analogous with a simple substitution reaction such as the H–D reaction:

$$H' + HD \rightarrow H'D + H \tag{18.18}$$

There are, of course, some important differences. In the case of the kink, the reaction coordinate is not simply a line along which displacements occur. Instead, it is a surface along which shear strains occur in a (110) direction. Furthermore, at a kink, the surface does not lie in a simply connected space. A circuit taken around the dislocation kink does not close. The closure mismatch for one circuit is the Burgers vector, b. For n circuits, it is nb.

If the atoms on the top of the glide plane are labeled T, and those on the bottom B, the reaction when the kink moves one step forward can be written:

$$B_1T + B_2 \rightarrow (B_1TB_2)^* \rightarrow B_1 + TB_2 \tag{18.19}$$

where T slides over B_1 to form the transition complex $(B_1TB_2)^*$. This complex then shears until T has its new partner, B_2. During this reaction the shear strain energy (proportional to the square of the shear strain) starts low, rises to a maximum, and then declines to zero. If the forces that arise during the process are defined by continuous potentials, the net resistance to the movement of the kink is relatively small. This is because, in this case, the variation of the energy with position is small and smooth. Accordingly, the force tending to move the kink toward, or away from, the $x = w/2$ position would pass through zero at this position, and would rise to a maximum at a position on one side or the other of the symmetric position.

Following Burdett and Price (1982), a correlation (Walsh) diagram for the "chemical reaction" just outlined is drawn (Figure 18.17). This shows graphically that the reaction is "disconcerted" in the Woodward–Hoffmann sense, and therefore thermally forbidden. The

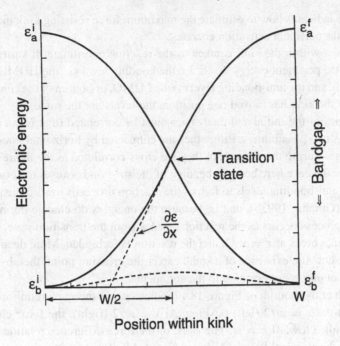

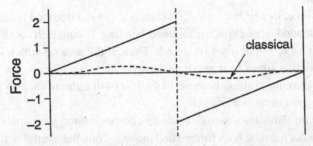

Figure 18.17 Walsh correlation diagram for kink motion. The width of the kink is w, and the reaction coordinate is x. The change of local symmetry at the $x = w/2$ position requires that the initial bonding state be correlated with the final anti-bonding state and vice versa. The diagram shows the correlation lines that connect the initial (superscript i) bonding (subscript b), and final (superscript f) anti-bonding (subscript a) states. The correlation lines are approximated by segments of parabolas. As the kink moves, bond bending causes the initial bonding level to rise with x^2 up to the transition state at $x = w/2$. Then it decreases until it reaches the final bonding state at $x = w = b$. The force $\partial\epsilon/\partial x$ increases linearly until $x = w/2$, and then changes sign discontinuously. The difference between the anti-bonding and the bonding energy levels is the LUMO–HOMO gap. At the transition-state singularity, a small gap opens (not shown). The LUMO–HOMO gap vanishes at the transition state, so the unpaired electron delocalizes.

diagram also indicates how to estimate the maximum force resisting kink motion, and how to calculate the pertinent activation energies.

The position within the kink is taken as the reaction coordinate. It varies from zero to w. Initially, the pertinent energy levels are the bonding level ϵ_b (the HOMO, or top of the valence band), and the anti-bonding level ϵ_a (the LUMO, or bottom of the conduction band). Finally, after the kink has moved one position, the levels are the same.

The energies of the initial and final states must be correlated (that is not arbitrary), and there are only two possibilities. Either they are connected by horizontal lines indicating no energy changes during the process, or they are cross-correlated as the figure indicates. The latter must be the case here because bending of the sp³-bonds causes the bonding levels to rise, and the anti-bonding levels to fall, as the reaction proceeds toward the transition state at $x = w/2$ (Gilman, 1993b), that is, because the energies do change during the process. The reverse process occurs as the reaction recedes from the transition state. Therefore, the correlation lines cross at $x = w/2$, and the reaction is forbidden. More detailed considerations would show the existence of a small gap at the crossing point, thereby removing the singularity, but this is not essential to the argument.

The sketch at the bottom of Figure 18.17 indicates that the maximum force on the kink is reached at the $x = w/2$ (left) position. At $x = w/2$ (right), the force changes sign so the kink is pulled toward $x = w$. The slope of the force difference relation is much more steep than in the classical Peierls–Nabarro theory. All of the quantitative parameters of the diagram are known.

The maximum force, $\partial\epsilon/\partial x$ at $x = w/2$, is sketched (dotted line) in Figure 18.17 for the parabolic energy dependences expected for bond bending. It equals $2(\Delta\epsilon/b)$ where $\Delta\epsilon$ is the energy gap ($\Delta\epsilon = \epsilon_a - \epsilon_b$); if we let $w = b$. Then, if the area on which it acts is taken to be $b^2/2$, the resistive stress is approximately $4(\Delta\epsilon/b^3)$. For silicon, $\Delta\epsilon = 1.2\,\text{eV}$ and $b = 3.76\,\text{Å}$ so the maximum stress is about 14.4 GPa (1440 kg/mm²) which is comparable with indentation hardness measurements.

The Walsh diagram also allows us to estimate the activation energy for dislocation motion. For this process, kinks must be both formed and moved. Thus the overall activation energy is the sum of the formation and motion energies. Forming a pair of kinks means that two of them must be moved apart by at least one atomic distance which takes an amount of energy, E_{hl}, equal to the LUMO–HOMO gap. Moving one kink takes $E_{hl}/2$, but two kinks must be moved for net dislocation motion, which takes E_{hl}. Therefore, the overall activation energy is $2E_{hl}$ which is consistent with the data of Figure 18.12.

It is interesting that what has been called hardness for centuries in the physical sciences is now directly linked through E_{hl} to what chemists have begun, as a result of the efforts of R.G. Pearson, to call "chemical hardness" (Pearson, 1997).

Some secondary effects in the motion of dislocations in covalently bonded crystals are:

(1) doping to form n- or p-type crystals,
(2) polarity in III–V compounds,
(3) photoplasticity.

In the context of the present model, doping simply changes the HOMO–LUMO gap in the vicinity of the dopant atoms. The polarity effect (difference in mobility of dislocations with half-planes ending on anions versus those with half-planes ending on cations) is related to the piezoelectric effect which is strong in these non-centrosymmetric crystals. This creates local electric fields which alter the HOMO and LUMO energy levels. In the photoplastic effect, electrons are excited across the HOMO–LUMO gap by incident photons, thereby reducing, or eliminating, the need for thermal excitation.

18.6.2 Temperature dependence

At high temperatures, the barrier to dislocation motion is the activation energy discussed above. This determines how rapidly the hardness declines with increasing temperature above the Debye temperature. The characteristic hardness versus temperature curves were first determined by Trefilov and Mil'man (1964). Figure 18.18 shows their data for Ge and Si. At low temperatures, two cases arise depending on whether the stress is compressive or tensile. In tension, the dislocation mobility is expected to decrease with decreasing temperature, and the hardness to continue to rise until fracture intervenes. In compression (including indentation hardness), shear-induced transformation to the structure of β-tin is encountered. The β-tin structure is metallic and ductile (Pharr, Oliver, and Harding, 1991), so dislocation mobility is high in this phase. Therefore, in this plateau region, as pointed out by Trefilov and Mil'man (1964), the hardness value depends on the stress needed to initiate the transformation, so the hardness becomes nearly independent of temperature.

Because of its technical importance, silicon has been chosen as the prototype for the discussion of temperature dependence, but quite similar arguments apply to other homopolar covalent crystals.

There is no direct way to measure kink velocities as a function of temperature. However, kink concentrations vary between 0 and 1, so the average is $1/2$, and most of the values are expected to lie between $1/4$ and $3/4$. Choosing $1/2$ for c_k, the dislocation velocities will be proportional to the kink velocities, and the hardnesses will also be proportional to the kink velocities.

It is convenient to normalize the hardness before examining the temperature dependence, that is, to divide the $H(T)$ values by H_0 the hardness at $T = 0$ K. This minimizes complications associated with the complex geometries of hardness indentations, with variable deformation rates, and so forth. Calculated and measured values for Si are shown in Figure 18.19. The measured values are those of J.H. Westbrook. The transition between the flat plateau at low temperatures and the exponential decrease at high temperatures occurs at the Debye temperature of Si (645 K).

The calculated points in Figure 18.19 are based on the idea that the kink concentration controls the temperature dependence since the process for creating kinks is essentially the same as the process for moving them.

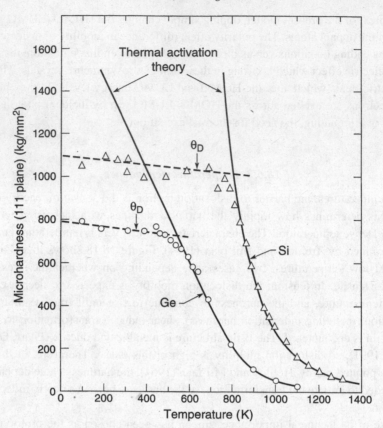

Figure 18.18 Data of Trefilov and Mil'man (1964) showing the temperature dependence of the indentation hardnesses of Ge and Si. Above the Debye temperatures, the hardnesses are determined by dislocation motion. Below it, the hardness is determined by the critical compressive stress for transformation from the diamond to the β-tin structure. Cracking also occurs at temperatures below the Debye temperature, causing scatter of the measured values.

There are just two kinds of kinks ("left" and "right") so they obey Fermi–Dirac statistics. Then, if the free energy of formation of a kink is:

$$G_k = E_k - T S_k \tag{18.20}$$

where E_k and S_k are the enthalpy and entropy of formation, respectively, the equilibrium concentration of kinks, c_k, at a temperature T is:

$$c_k = \{1 + \exp[(E_k - T S_k)/kT]\}^{-1} \tag{18.21}$$

Thus, c_k goes to zero as T becomes small, and tends toward unity as T becomes large, provided the kink entropy S_k is substantial.

The normalized hardness number is:

$$H/H_o = 1 - c_k \tag{18.22}$$

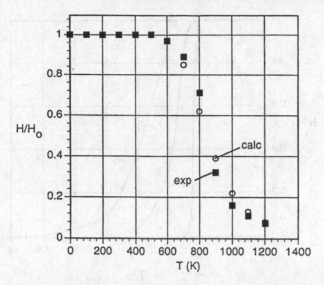

Figure 18.19 Comparison of theory and experiment for the temperature dependence of the hardness of silicon. The parameters of the calculated points are as follows: the normalizing hardness $H_o = 730\,\text{kg/mm}^2$; the entropy parameter $A = 2.7 \times 10^{-4}$; the enthalpy $E_k = 0.6\,\text{eV}$. The "experimental" points are from a smoothed line drawn through data of J.H. Westbrook.

Plastic flow is assumed to occur when $c_k \approx 1/2$. Then:

$$H/H_o = 1 - [1 + A\exp(6977/T)]^{-1} \qquad (18.23)$$

where $E_k = E_g/2 = 0.6\,\text{eV}$, and $A = \exp(-S_k/k)$.

Figure 18.19 shows that this fits the experimental values very well with $A = 2.7 \times 10^{-4}$. This yields a kink entropy of $S_k = 7.1 \times 10^{-4}\,\text{eV/K}, \approx 8$ entropy units. Since the average coordination number at a kink is about 8, this is reasonable. Also, at the Debye temperature, $\Theta_D = 645\,\text{K}$, $S_k\Theta_D/E_k \approx 0.8$ as might be expected.

$$H/H_o = 1 - c_k$$
$$= 1 - [1 + A\exp(E_k/kT)]^{-1} \qquad (18.24)$$

Since the break in the H versus T curve for Si occurs near $T = \Theta_D$, it is of interest to plot the hardness data for other homopolar crystals in terms of the temperature relative to the Debye temperature, or $T_{red} = T/\Theta_D$. The results for diamond ($\Theta_D = 1800\,\text{K}$), germanium ($\Theta_D = 374\,\text{K}$), and silicon ($\Theta_D = 645\,\text{K}$) are shown in Figure 18.20. In all three cases the break occurs near Θ_D. Also, the rates of exponential decrease are ordered, as expected, in terms of the glide activation energies.

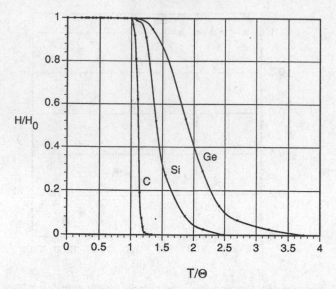

Figure 18.20   Plot of reduced hardness versus reduced temperature for three homopolar crystals. The Debye temperatures that are used for C, Si, and Ge are 1800, 645, and 374 K, respectively.

18.6.3 Chemical stability

Since dislocation mobility and chemical reactivity are connected, understanding the former involves the latter. Chemical stability (inverse reactivity) is determined by chemical "hardness", and the principle of maximum hardness (Pearson, 1997). The chemical hardness is determined by the size of the gap in the bonding energy spectrum of molecules and solids (Parr and Yang, 1989). It is defined to be equal to one-half the gap (i.e., the difference between the lowest unoccupied molecular orbital (LUMO), and the highest occupied molecular orbital (HOMO)).

Hardness (scratch or indentation) in the physical sciences (geology, metallurgy, ceramics, etc.), and the chemical "hardness" that came from studies of the chemical behavior of acids and bases, have the same fundamental basis: the gap in the bonding energy spectrum. Physical hardness through its dependence on dislocation mobility is also determined by the gap in the bonding energy spectrum. The gap determines stability (Burdett, 1995), and therefore determines indentation hardness for covalent crystals as shown for various semiconductors in Figure 18.21. The members of each of the sets of crystals are isoelectronic. The hardest set (Group IV) is homopolar having the most localized bonding. The set with the most ionicity (II–VI compounds) is the softest and has the most delocalized bonding. The set of III–V compounds lies in between.

18.7 Molecular solids

Molecular crystals and glassy polymers form a fourth class of solids. They are bound internally by two kinds of bonding, intramolecular and intermolecular interactions. The

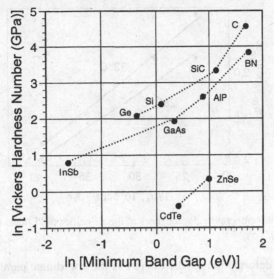

Figure 18.21 Indentation hardness versus minimum band gaps for crystals with tetrahedral isoelectronic bonding (Groups II–VI, III–V, and IV).

internal (intramolecular) bonding is usually covalent (localized), while the bonding between molecules is of the van der Waals (London) type. In the covalent case, the atoms bond by exchanging charged electrons. In the van der Waals case, the molecules bond by exchanging neutral photons (electromagnetic quanta), and the bonding is much weaker but of much longer range. When the range is large enough for the transmission time between molecules to exceed the vibrational periods of the molecular dipoles, the forces are called Casimir forces (or retarded forces).

Molecular solids have two limiting structures with a vast variety of intermediate cases. One limit consists of crystals of small spherically symmetric neutral particles, for example, the noble gases, and small four-fold molecules such as methane and carbon tetrachloride. These crystals have close-packed arrays of particles, and are very soft, as well as being plastically nearly isotropic.

The other limit consists of high molecular weight linear polymers, such as high density polyethylene. If the polymer molecules in these are oriented to lie parallel, the resulting materials are fibrous, and therefore very anisotropic. Dislocations can move readily along the fibers, but only with great difficulty transverse to them.

A technique for directly observing dislocations in polymers has yet to be devised. However, some information about their behavior can be obtained indirectly by observing the motion of "Lüders bands" along stressed fibers that are initially unoriented. These bands are the relatively sharp junctions that form between unoriented and oriented sections of stressed fibers. At constant shape they propagate along the lengths of fibers, thereby converting a specimen from the unoriented to the fully oriented state.

Since the shape of such a front is constant as it propagates, to preserve continuity there is a relation between the average deformation rate, $\partial \delta / \partial t$, the velocity of the band's front, v_B,

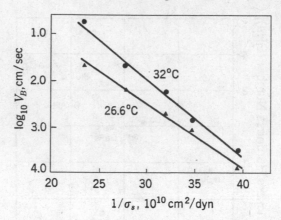

Figure 18.22 Stress and temperature dependences of the velocities of "Lüders band" fronts in Nylon 6-10 (after Dey, 1967).

and the gradient of deformation $\partial\delta/\partial l$, where l is the coordinate parallel to the fiber axis (Hahn, 1962):

$$\partial\delta/\partial t = v_B(\partial\delta/\partial l) \tag{18.25}$$

Thus an expression for the deformation rate, such as the Orowan equation, can be converted into an expression for the gradient. Integration of the latter gives the shape of the front. For small deformations (approximately constant stress across the front) the shape of the front will be the same as that of a creep curve with time replaced by position.

If a band front shape is constant, the mobile dislocation density at each point within it is constant, although the details of the distribution are not known. This allows relative, average, dislocation velocities to be measured. The average velocity at a particular cross-section (where $\delta = \delta^*$) is given by:

$$\langle v_d \rangle = Zv_B(\partial\delta/\partial l)_{\delta^*} \tag{18.26}$$

where Z is a constant. Dey (1967) has applied this method to Nylon 6-10 filaments. Some of his data are shown in Figure 18.22. They show the temperature dependence of the "dislocation" motion between the molecules in a polymer filament, and the high sensitivity of the motion to applied stress. Note that the velocity follows the form of Equation (18.5).

18.8 Alloys and intermetallic compounds

This is the largest category of materials by far. For practical materials, there are about 50 metallic elements that can be produced without prohibitive cost. The binary combinations of these alone come to 1225; and there may be several concentrations of interest in each binary system. In addition, there are 13 200 ternary combinations. Thus the overall size of this category is immense.

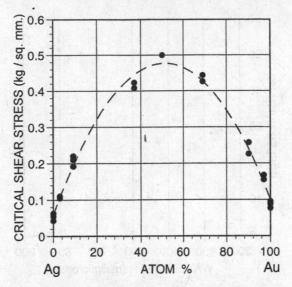

Figure 18.23 Effect of forming a simple alloy of Ag and Au on the critical shear stress for yielding. Alloying changes the stress needed for dislocation motion in this system by a factor of about ten, although the yield stress remains relatively low.

In mixtures of metal atoms, it is apparent that the volumetric density of valence electrons must vary from one atom to another, hence, from one atomic position to another along the core of a dislocation. Since the energy fluctuates from one atomic position to another, so do the energy gradients, or forces. As a result, it is to be expected that the resistance to dislocation motion will increase with the concentrations of the fluctuations.

As a simple example, consider a binary alloy, AB. Starting with pure A, let c be the fractional concentration of B, so the concentration of A will be $1 - c$. Then the average number of energy fluctuations will be proportional to $c(1 - c)$. This is zero when $c = 0$ or 1; and it reaches a maximum of 0.25 when $c = 0.5$. This is just what is observed (Figure 18.23) in some simple alloy systems such as Ag–Au (Sachs and Weerts, 1930). The behavior is much more complex in many cases, but the Ag–Au case is particularly simple because the atoms have very nearly the same size and are isoelectronic. However, their bulk moduli are quite different (Au is 1.7 times stiffer than Ag), indicating that their valence electron concentrations differ in a similar ratio. In Figure 18.23 the maximum increase in the shear yield stress is about 4×10^{-5} Mbar, while the average shear modulus is about 0.8 Mbar, so the maximum internal stress fluctuations are about 5×10^{-5}. The maximum hardening effect is a factor of about ten according to Figure 18.23 which confirms how sensitive dislocation mobilities are to small deviations from homogeneity.

In intermetallic compounds, the situation is still more complicated since the electron density changes not only from one atomic site to another, but also from one interface (bond) between two sites to another. Then, if the symmetry of the crystal structure is low, the pattern

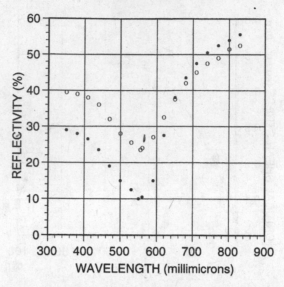

Figure 18.24 Effect of plastic deformation on the color of $AuAl_2$ crystals. The solid circles represent crystals annealed at 500 °C, while the open circles are for crystals deformed 81% in compression. The color change is from purple to grayish blue.

of fluctuations can become quite complex. In principle, however, the dislocation mobility is determined by the pattern and intensities of the electron density fluctuations.

Evidence that the effects are directly electronic in origin, rather than indirectly mechanical, is provided by the fact that deformation often changes the color of a compound. This effect is quite large, for example, in gold aluminide (Figure 18.24). This bright, purple metallic compound acquires a grayish cast when it is deformed (Shih, 1963).

18.9 Oxide crystals (including silicates)

There are several classes of oxides, depending on the valences of the cations. The first is the alkaline earth oxides (oxides of Be, Mg, Ca, Sr, and Ba) in which the chemical bonding is ionic and the mechanical behavior is similar to that of the alkali halides. Next are the oxides with trivalent cations such as B, Al, and Ga in which the bonding is a mixture of ionic and covalent and in which dislocation mobility is low. After these come the quadrivalent glass-forming elements, Si, Ge, Sn, and Pb. These oxides crystallize, but slowly, so it is easy to cool them fast enough to bypass crystallization, and obtain glasses. Then there are a host of transition metal oxides, as well as mixed cation glasses such as crysoberyls, garnets, and spinels. In garnets, with their complex crystal structures, dislocation mobilities tend to be particularly low. This makes them the most refractory of all known oxide materials with low dislocation mobilities at the very highest temperatures.

18.10 Glasses

Another category of materials is that of glasses, including insulators (e.g., silicate glasses), semiconductors (e.g., chalcogenide glasses), and metals (e.g., metal–metalloid glasses). In glasses the structures are non-periodic, so the energies of dislocations in them depend markedly on the exact positions of the cores. In addition, covalent bonds form between the metal and metalloid atoms, and these inhibit dislocation motion.

In metallic glasses, dislocation mobilities tend to be very low because of their lack of periodicity. These glasses tend to be elastic up to high stresses (of order $G/2\pi = 0.16$) where they yield in compression, and often break in tension. Thus they are nearly ideally elastic–plastic. The deformation in them is heterogeneous, and rapid as shown by the fact that they "cry" during yielding; and a multitude of "glide bands" are seen on their surfaces after yielding.

Hardness values for a number of metallic glasses are plotted in Figure 18.25 against their shear moduli, with the yield stress $\approx H/3 \approx G/6\pi$. The glasses exhibit some ductility because dislocations have very large multiplication rates in them, as indicated by the large amounts of acoustic emission emitted during yielding. The hardness numbers for glasses are large fractions of their shear moduli according to the data (Figure 18.25) of Chou, Davis, and Narasimhan (1977). The proportionality coefficient, 0.16, is close to the value found for Group IV elements of 0.17 (Gilman, 1968b). Both of these values are close to $1/2\pi$.

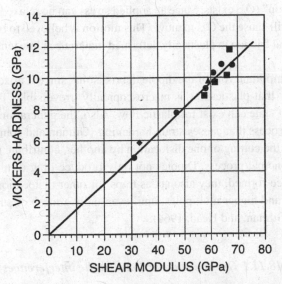

Figure 18.25 Hardnesses of metallic glasses as a function their shear moduli: ♦ Fe–B, ▲ Fe(40)Ni (38)Mo(4)B(18), ■ Cu(68)Zr(32), ● other glasses; slope 0.16. Data from Chou, Davis, and Narasimhan (1977). The line through the data has the equation $H = G/2\pi$.

A simple theory for the resistance to dislocation motion in a glass is that the lack of periodicity requires that, during shearing at the cores, atoms at the top of the glide plane must ride up and over those on the bottom, and this creates a center of dilatation. It is known from elasticity theory, for such a center, that if the dilatational displacement is δ, the strain energy is about $8\pi Gb\delta^2$. Equating this to the work done by the applied shear stress during the process, τb^3, the required stress is $\tau = 8\pi G(\delta/b)^2 = 8\pi G\epsilon^2$, where ϵ is the local strain. From plasticity theory, $\tau = H/6$, so the implied strain is about 3%, a reasonable value.

18.11 Self-interactions (strain hardening)

An important feature of dislocation motion is its three-dimensionality. This results from the cylindrical symmetry of the screw orientations. Because of this symmetry they can cross-glide from their primary glide planes to secondary planes, and back to the primary one. This facilitates rapid multiplication of their lengths, and also results in "dipole debris" that is left in their wakes as they move (Johnston and Gilman, 1960). The dipoles consist of pairs of edge dislocations within which the interactions of the stress fields are strong.

If the applied stress is zero, there are two orientations for dipoles of equal energy. That is, the line connecting the two dislocations, and perpendicular to both of them, can lie at either $+45°$ or at $-45°$ with respect to the normal to the glide plane. Furthermore, a given length of dipole can contain more than one orientation. Between each pair of orientations an "orientation junction" (OJ) exists. Since an applied stress can be relaxed by favoring one OJ over the other, it will cause the OJs to move. This motion is believed to be the mechanism of the Bordoni internal friction in plastically deformed materials at low temperatures (Gilman, 1996a).

Another more important role of the dipole debris is that since it arises from a stochastic process, it ensures that plastic flow is macroscopically irreversible. This, in turn, means that no equation of state can exist for plastic flow. Also, the generation of dipole debris is the fundamental process that causes strain hardening (Gilman and Johnston, 1960). Other processes, such as the cutting of one dislocation by another, contribute to strain hardening, but are not the principal process. Dipoles not only produce drag on the dislocations that form them, but once formed, they also act as traps for other dislocations to form tripoles, then quadrupoles, and higher multipoles, until eventually a mesh of sub-grain boundaries is formed (Chen, Gilman, and Head, 1964).

18.11.1 Dynamic interactions (traffic interferences)

As a group of dislocations moves, the individual dislocations within the group interact with one another in a manner similar to the interactions within a group of automobiles moving along a roadway (Gilman, 1968c). In both cases, the responses of the individuals to the driving forces are not instantaneous, so the interactions lead to decreases in the average velocities. The response delays result from both inertial and viscous drag effects. As the density of lines in the group increases, the average velocity of the group decreases, slowly at first and then quickly, as a critical density is approached where the velocity drops to zero. To restart the array, the applied stress must increase. Then "waves" of decreased and increased density travel backward and forward through the array until it reaches a steady state, or halts again. The result is a serrated stress–strain curve.

There are other manifestations of the collective effects, but they will not be discussed here.

18.12 Activation of motion

There are a variety of barriers to dislocation motions, both intrinsic and extrinsic, as indicated in the previous sections. These barriers can be overcome in three general ways:

(1) by passing over the barrier,
(2) by tunneling through the barrier,
(3) by reducing the height of the barrier.

Since dislocations are not point particles, but are lines, they have more constraints, as well as more degrees of freedom, associated with their motion than do particles. Thus the direct, or "brute", path leading to motion is not always the most effective one.

In surmounting a barrier, there is much to be learned from the history of high-jumping. In the beginning, high-jumpers simply ran up to the pole with their chests parallel to the pole and jumped. Then, a clever one discovered that he could get over a higher pole if he approached it with his chest perpendicular to the pole, and "scissored" his legs thereby lowering the height to which he needed to raise his center of gravity. Next, someone discovered the "western-roll" in which the scissoring action is combined with a face-down rolling action. Finally, it was found that a slight advantage could be had by approaching the pole with the back facing parallel to it rather than the chest. The difference in jump height between the least, and the most, effective techniques has been large, perhaps as much as 30%.

Dislocations also take advantage of their elongated forms, and their flexibilities in overcoming barriers. This is why they do not follow conventional statistical mechanics in their behavior, and why their "constitutive relations" are unusual. They are not particles.

Shear stress drives dislocation motion. It is sometimes assisted by thermal fluctuations, and sometimes hindered by them. At temperatures below the Debye temperature where the phonon wavelengths are long compared with atomic diameters, coupling of thermal fluctuations to sharp kinks and localized traps is poor. In this regime, the thermal energy density also decreases. Because of both of these effects, stress plays a more important role than thermal fluctuations in activating motion. When the stress is sufficiently high, at localized barriers, it can cause tunneling through a barrier.

In addition to helping dislocations to surmount barriers, and to tunnel through them, shear stress may reduce barrier heights by affecting the local bonding electrons. Covalent bonding opens up LUMO–HOMO gaps in the local bonding energy spectrum. However, shear strains close the gaps by raising the HOMO levels, and lowering the LUMO levels, thereby weakening the bonds (Gilman, 1995a, 1996b).

The phenomenology of plastic deformation is exceedingly complex. This has the advantage of allowing the behavior of individual materials to be tailored to the needs of particular engineering applications, but it also makes it impossible to devise general theoretical descriptions. As the preceding paragraphs suggest, the issue of motion activation is complicated with several limiting cases depending on whether intrinsic, or extrinsic, mechanisms are involved. More than one mechanism can be operative at one time, causing still more complication.

In crystals with non-local bonding where dislocation lines move concertedly, the motion of the lines is not activated. It is resisted by the viscous drag created by nearly free electrons, and by phonons. As described above, these drag effects are relatively small.

18.12.1 Temperature activation

Unlike simple chemical reactions in dilute gases, or dilute solutions, where independent particles are the participants and Boltzmann statistics apply, the reactants in the case of dislocations are not independent. If a reaction occurs at some place along a dislocation line, the probability that the next reaction will occur adjacent to the current one is much higher than that it will occur at a random position along the line. Thus the reaction events are highly correlated, and the atoms involved are not statistically independent. Hence, Boltzmann statistics in a simple form should not be expected to apply.

When kinks are important, since they have only two varieties (+ and −), Fermi–Dirac statistics operate. This is why such systems (e.g., pure silicon) have no temperature dependence for plastic flow at low temperatures (Gilman, 1995b).

At high temperatures (i.e., above Θ_D), the behavior is usually described by the "Becker model", originally due to the German physicist, R. Becker (1925). He proposed that the effect of stress on the rates of chemical reactions may be described by applying a correction to the Arrhenius function. The correction consists of subtracting the work, σV, done by the stress during the reaction from the activation energy, Q. The work done is the stress times an area, giving a force, times a displacement which yields a reaction volume, the "activation volume" V. Thus the effective activation energy Q' becomes $Q' = Q - \sigma V$. Unfortunately for the theory, this often does not give the observed temperature dependence.

In the case of a dislocation (which is two-dimensional), the line sweeps out an area (the glide area) so that two line displacements are involved, one parallel and one perpendicular to the direction of motion of the line. Now the shape of the activated complex must be optimized with respect to the two displacements, not just one. When this is done, the effective activation energy, Q'', becomes $Q'' = Q/\xi\sigma$ where ξ is a coefficient (Gilman, 1965). Unfortunately, like the Becker theory, this also does not give the observed temperature dependence.

Neither of the expressions, for Q' or Q'', is consistent with observations of the temperature dependences of the yield stresses of materials in which dislocation mobility is limited by extrinsic factors. However, if they are combined into one Arrhenius-type function, rate $\sim \exp[-(Q'Q'')/kT] \sim \exp[-(Q - \sigma V)/\xi \sigma T]$, and it is assumed that yielding occurs when the rate reaches a critical value where the argument of the exponential function equals a constant, say ϕ, then the yield stress, σ_y, after choosing a new constant, $A = \phi \xi$, is given by:

$$\sigma_y = Q/(V + AT) \tag{18.27}$$

which is approximately correct in most cases. When the temperature approaches zero, the yield stress becomes constant; and at higher temperatures it decreases inversely with the temperature.

18.12.2 Stress activation

At low temperatures, the stress needed to move dislocations often becomes quite high so the applied strain energy per atom becomes large compared with the thermal energy per atom. For this case, it makes little sense to say that dislocation motion is "thermally activated". For example, consider a material with a shear modulus $G = 120\,\mathrm{GPa}$, Burgers displacement $b = 2.5\,\text{Å}$, the atomic mass of iron, $56\,\mathrm{amu}$, and let the temperature be that of liquid nitrogen, $80\,\mathrm{K}$. Then, if the applied stress is $G/100$, the strain energy per atom is about $100\,\mathrm{meV}$, the thermal energy $kT/2$ is only about $3.4\,\mathrm{meV}$, while the zero-point energy is about $32\,\mathrm{meV}$. Therefore, the driving force provided by the stress is 30 times the thermal energy, so the assistance given to the stress by the thermal energy is negligible. At still lower temperatures, thermal energy helps even less.

The mechanism of activation caused purely by stress appears to be the quantum tunneling of bonding electrons into anti-bonding states. Equivalently, the shear stress (strain) may close the LUMO–HOMO gap, thereby eliminating the activation barrier. For tunneling the expression for the average dislocation velocity, v, takes the form (see also Equation (18.5)):

$$v = v_{max}[\exp(-D/\tau)] \tag{18.28}$$

where v_{max} is near the speed of elastic shear waves, and D is a drag coefficient. This represents the observed behavior for a large number of cases. The velocity is averaged over "stick" periods where $v_i \approx 0$, and "slip" periods where $v_i \approx v_{max}$, with v_i being the instantaneous velocity.

References

Alexander, H. and Haasen, P. (1968). Dislocations and plastic flow in the diamond structure, *Solid State Phys.*, **22**, 28.

Alexander, H. (1979). Models of dislocation structure, Proc. Int. Symp. on Dislocations in Tetrahedrally Coordinated Semiconductors, *J. Phys. Coll.* C6, **40**, C6-1.

Andrade, E.N. da C. (1934). *Philos. Mag.*, **17**, 497, 698.

Becker, R. (1925). *Phys. Z.*, **26**, 919.

Burdett, J.K. and Price, S.L. (1982). Application of Woodward–Hoffmann ideas to solid-state polymorphic phase transitions with specific reference to polymerization of S_2N_2 and the black phosphorous to A7 (arsenic) structural transformation, *Phys. Rev. B*, **25**, 5778.

Burdett, J.K. (1995). *Chemical Bonding in Solids*, p. 298. New York: Oxford University Press.

Chen, H.S., Gilman, J.J., and Head, A.K. (1964). Dislocation multipoles and their role in strain-hardening, *J. Appl. Phys.*, **35**, 2502.

Chou, C.P., Davis, L.A., and Narasimhan, M.C. (1977). Elastic constants of metallic glasses, *Scr. Metall.*, **11**, 417.

Cohen, M.L. (1991). *Philos. Trans. R. Soc. London, Ser. A*, **334**, 501.

Cottrell, A.H. (1953). *Dislocations and Plastic Flow in Crystals*, p. 74. Oxford: Oxford University Press.

Coulson, C.A. (1952). *Valence*, p. 71. Oxford: Clarendon Press.

Dey, B.N. (1967). *J. Appl. Phys.*, **38**, 4144.

Duesbery, M.S. and Richardson, G.Y. (1991). The dislocation core in crystalline materials, *Solid State Mater. Sci.*, **17**, 1.

Foreman, A.J., Jaswon, M.A., and Wood, J.K. (1951). *Proc. Phys. Soc. A*, **64**, 156.

Frank, F.C. (1992). Body-centered cubic: a close-packed structure, *Philos. Mag. Lett.*, **66**, 81.

Galligan, J.M., Lin, T.H., and Pang, C.S. (1977). Electron–dislocation interaction in copper, *Phys. Rev. Lett.*, **38**, 405.

Galligan, J.M., Goldman, P.D., Motowildo, L., and Pellegrino, J. (1986). *J. Appl. Phys.*, **59**, 3747.

Gilman, J.J. (1956). Plastic anisotropy of zinc monocrystals, *J. Met.*, October, 2.

Gilman, J.J. (1959). Plastic anisotropy of LiF and other rocksalt-type crystals, *Acta Metall.*, **7**, 608.

Gilman, J.J. (1960). The plastic resistance of crystals, *Aust. J. Phys.*, **13**, 327.

Gilman, J.J. and Johnston, W.G. (1960). Behavior of individual dislocations in strain-hardened LiF crystals, *J. Appl. Phys.*, **31**, 687.

Gilman, J.J. (1961). Prismatic glide in cadmium, *Trans. Metall. Soc. AIME*, **221**, 456.

Gilman, J.J. and Johnston, W.G. (1962). Dislocations in lithium fluoride crystals. In *Solid State Physics*, ed. F. Seitz and D. Turnbull, Volume 13, p. 147. New York: Academic Press.

Gilman, J.J. (1965). Dislocation mobility in crystals, *J. Appl. Phys.*, **36** (10), 3195.

Gilman, J.J. (1968a). Dislocation motion in a viscous medium, *Phys. Rev. Lett.*, **20**, 157.

Gilman, J.J. (1968b). Escape of dislocations from bound states by tunneling, *J. Appl. Phys.*, **39**, 6068.

Gilman, J.J. (1968c). The plastic response of solids. In *Dislocation Dynamics*, ed. A.R. Rosenfeld, G.T. Hahn, A.L. Bement, and R.I. Jaffee, p. 3. New York: McGraw-Hill.

Gilman, J.J. (1970). Hardnesses of carbides and other refractory hard metals, *J. Appl. Phys.*, **41** (4), 1664.

Gilman, J.J. (1973). Hardness of pure alkali halides, *J. Appl. Phys.*, **44** (3), 982.

Gilman, J.J. (1975). Flow of covalent solids at low temperatures, *J. Appl. Phys.*, **46**, 5110.

Gilman, J.J. (1993a). Shear-induced metallization, *Philos. Mag. B*, **67**, 207.

Gilman, J.J. (1993b). Why silicon is hard, *Science*, **261**, 1436.

Gilman, J.J. (1995a). Mechanism of shear-induced metallization, *Czech. J. Phys.*, **45**, 913.

Gilman, J.J. (1995b). Quantized dislocation mobility. In *Micromechanics of Advanced Materials*, ed. S.N.G. Chu, P.K. Liaw, R.J. Arsenault, K. Sadananda, K.S. Chan, W.W. Gerberich, C.C. Chau, and T.M. Kung, p. 3. Warrendale, PA: The Minerals, Metals, & Materials Society.

Gilman, J.J. (1996a). Bordoni internal friction via dislocation dipole orientation junctions. In *The Johannes Weertman Symposium*, ed. by R.J. Arsenault, D. Cole, T. Gross, C. Kostorz, P. Liaw, S. Parameswaran, and H. Sizek, p. 411. Pittsburgh, PA: The Minerals, Metals & Materials Society.

Gilman, J.J. (1996b). Mechanochemistry, *Science*, **274**, 65.

Greenman, W.F., Vreeland, T., and Wood, D.S. (1967). *J. Appl. Phys.*, **38**, 4011.

Grimvall, G. and Thiessen, M. (1986). The strength of interatomic forces, *Science of Hard Materials, Inst. Phys. Conf. Ser. No. 75*, p. 61. Bristol: Adam Hilger.

Hahn, G.L. (1962). *Acta Metall.*, **10**, 727.

Hirth, J.P. and Lothe, J. (1968). *The Theory of Dislocations*. New York: McGraw-Hill.

Johnston, W.G. and Gilman, J.J. (1960). Dislocation multiplication in lithium fluoride crystals, *J. Appl. Phys.*, **31**, 632.

Johnston, W.G. (1962). Effects of impurities on the flow stress of LiF crystals, *J. Appl. Phys.*, **33**, 2050.

Kobelev, N.P., Soifer, Ya.M., and Al'shits, V.I. (1971). Relationship between viscous and relaxation components of the dislocation attenuation of high-frequency ultrasound in copper, *Sov. Phys. Solid State*, **21** (4), 680.

Kojima, H. and Suzuki, T. (1968). Electron drag and flow stress in niobium and lead at 4.2 K, *Phys. Rev. Lett.*, **21**, 896.

Kveder, V., Omling, P., Grimmeiss, H.G., and Osipyan, Yu. A. (1991). Optically detected magnetic resonance of dislocations in silicon, *Phys. Rev. B*, **43**, 6569.

Ledbetter, H. (1991). Atomic frequency and elastic constants, *Z. Metallkd.*, **82** (11), 820.

Li, J.C.M. (2000). Charged dislocations and the plasto-electric effect in ionic crystals, *Mater. Sci. Eng.*, **287A**, 265.

Love, A.E.H. (1944). *A Treatise on the Mathematical Theory of Elasticity*, 4th edn., p. 221. New York: Dover Publications. 1st edn., Cambridge University Press, Cambridge, 1892.

Mason, W.P. (1968). Dislocation drag mechanisms and their effects on dislocation velocities. In *Dislocation Dynamics*, ed. A.R. Rosenfeld, G.T. Hahn, A.L. Bement, and R.I. Jaffee, p. 487. New York: McGraw-Hill.

McColm, I.J. (1990). *Ceramic Hardness*, p. 93. London: Plenum Press.

Murr, L.E. (1975). *Interfacial Phenomena in Metals and Alloys*, p. 165, Reading MA: Addison-Wesley.

Nabarro, F.R.N. (1987). *Theory of Crystal Dislocations*, pp. 120 and 257ff. New York: Dover Publications.

Nabarro, F.R.N. (1997). Theoretical and experimental estimates of the Peierls stress, *Philos. Mag. A*, **75**, 703.

Nadgornyi, E. (1988). Dislocation dynamics and mechanical properties of crystals. In *Progress in Materials Science*, Volume 31, ed. J.W. Christian, P. Haasen, and T.B. Massalski, pp. 251–264. Oxford: Pergamon.

Orowan, E. (1934). *Z. Phys.*, **89**, 605, 614, 634.

Parr, R.G. and Yang, W. (1989). *Density Functional Theory of Atoms and Molecules*. New York: Oxford University Press.

Pearson, R.G. (1997). *Chemical Hardness*. New York: Wiley-VCH.

Peierls, R.E. (1940). *Proc. Phys. Soc.*, **52**, 34.

Pharr, G.M., Oliver, W.C., and Harding, D.S. (1991). New evidence for a pressure-induced phase transformation during the indentation of silicon, *J. Mater. Res.*, **6**, 1129.

Polanyi, M. (1934). *Z. Phys.*, **89**, 660.

Powers, R.W. (1954). Internal friction in solid solutions of oxygen–tantalum, *Acta Metall.*, **3**, 135. He "cooked" b.c.c. metal wires in ultrahigh vacuum for long periods, until their Snoek effect internal-friction peaks became very small. The wires were then very soft.

Price, D.L., and Cooper, B.R. (1989). *Phys. Rev. B*, **39**, 4945.

Price, D.L., Cooper, B.R., and Wills, J.M. (1992). *Phys. Rev. B*, **46**, 11368.

Reif, F. (1965). *Fundamentals of Statistical and Thermal Physics*. New York: McGraw-Hill.

Sachs, G. and Weerts, J. (1930). *Z. Phys.*, **62**, 473.

Shih, H.C. (1963). Color changes in the compound $AuAl_2$ induced by imperfections, *M.S. Thesis*, Metallurgical Engineering, University of Illinois, Urbana, IL.

Sleezwyk, A.W. (1963). *Philos. Mag.*, **8**, 1467.

Sumino, K. (1989). *Inst. Phys. Conf. Ser.*, **104**, 245.

Suzuki, H. (1968). Motion of dislocations in body-centered cubic crystals. In *Dislocation Dynamics*, ed. A.R. Rosenfeld, G.T. Hahn, A.L. Bement, and R.I. Jaffee, p. 679. New York: McGraw-Hill.

Taylor, G.I. (1934). The mechanism of plastic deformation of crystals. Part I. Theoretical, *Proc. R. Soc. London, Ser. A*, **145**, 362.

Teter, D.M. (1998). Computational alchemy: the search for new superhard materials, *Mater. Res. Soc. Bull.*, **23** (1), 22.

Trefilov, V.I. and Mil'man, Yu.V. (1964). Aspects of the plastic deformation of crystals with covalent bonds, *Sov. Phys. Dokl.*, **8**, 1240.

Vitek, V. (1974). *Cryst. Lattice Defects*, **5**, 1.

Volterra, V. (1907). Sur l'équilibre des corps élastiques multiplement connexes, Paris, *Ann. Ec. Norm. Ser. 3*, **24**, 401–517.

Woodward, R.B. and Hoffmann, R. (1970). *The Conservation of Orbital Symmetry*. Weinheim: Verlag Chemie.

Section VI

Fracture resistance

19

Mechanics of cracks

The ultimate determinant of strength is fracture. Its upper limit is the cohesive strength of a material, but it is rare that the cohesive strength can fully manifest itself. Almost always, mechanisms intervene that concentrate the macroscopic stress into small regions where the local stress may be 10 000 times, or more, larger than the nominal applied stress. The most effective stress concentrators are cracks. Like levers, they concentrate the work that is done by macroscopic applied forces into small microscopic volumes.

19.1 Elements of cracking

Two centuries ago young men could make a living by splitting trees lengthwise into "rails" to be used for fence construction. Cracking could also be used to split large rocks in quarries. Figure 19.1 illustrates a splitting crack. Here a crack has traversed about half the length of a rod of material (perhaps a wooden log). The crack bisects the thickness, $2t$, of the rod (whose width is w). The length of the crack is L, and it forms two cantilever beams each of thickness t and width w. Suppose that a wedge applies forces, F, pushing the ends of the cantilevers apart, thereby displacing each of them a distance, h, from the center-line of the rod. These displacements increase by small amounts, dh, if the tip of the crack advances incrementally by an amount dL.

The forces applied by the wedge do small amounts of work, $dw = F\,dh$, during an advance of the crack dL. It is assumed that the rod is elastic, and that the advance of the crack is reversible. Then, since the elastic deflection of a cantilever is dependent on its length, dh in terms of dL can be calculated. If L is large compared with t and w, the "elementary" theory of elastic beams is adequate for the calculation (Gillis and Gilman, 1964). When the crack advances by the amount dL it creates two new amounts of surface area, each of size $w\,dL$ for a total of $2w\,dL$. Therefore, if the specific surface energy is S, this requires an energy $2Sw\,dL$. In order to conserve the energy of the system, this must be supplied by the input work, $2F\,dh$, which must also supply the incremental increase in elastic energy associated with the increased lengths of the two cantilevers.

247

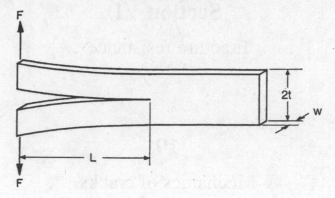

Figure 19.1 Splitting a rod as a symmetric double-cantilever.

Notice that the conditions needed to advance the crack can be calculated without calculating the local stress. All that is necessary is the formation of an energy (or work) balance. Recognition of this by A.A. Griffith (1920), and his experimental confirmation of it, was a great advance in the state of this subject. The details of the stress distribution need not be known. Indeed, the stress distribution at the tip of a crack is an invariant when the critical condition for propagation of the crack is met. However, the length of the crack must be known when the propagation condition is just satisfied. Thus, the critical condition for cracking cannot be obtained in terms of the properties of the material alone. It depends on the geometric configuration. In particular, it depends on a characteristic length of the system. Therefore, fracture phenomena are always inhomogeneous in the sense that they involve a mixture of geometric and material properties. This tends to complicate them.

In the "fracture mechanics" of engineering practice, for historical reasons, the geometric and material aspects of fracture phenomena are mixed into parameters called "stress-intensity factors". This practice had its origin in the false premise that cracks advance when a critical value of the local stress is reached. This has created an awkward body of knowledge regarding fracture mechanics, as well as awkward methods of analysis. These can be avoided by separating the two aspects of fracture so that more descriptive language can be used, and the powerful, traditional concept of energy, and its conservation, can be used. It is especially useful to be able to analyze dynamic fracture phenomena in terms of energies, rather than stresses. Then the dynamics can be described in terms of the kinetic energy, while the statics is described in terms of potential energy.

Fracture phenomena will be described here in terms of energies. Nomenclature is partly a matter of convention, but not entirely because the physical meaning of a "stress-intensity factor" is questionable. During the splitting of the slender rod of Figure 19.1, for example, the maximum stress at the tip of the crack is invariant for reversible movement of the crack; but it is indefinite (a singularity) within the framework of elasticity theory. This was clearly recognized by Griffith because he began his studies of fracture with studies of stress concentrators, and the magnitudes of concentrated stresses. Thus the question he faced was "if solids containing cracks have indefinite stresses at the crack tips, why do they not simply fall into pieces?". The answer, he discovered, is that high elastic stresses are a necessary, but not sufficient, condition for crack propagation.

For crack propagation, not only must the local linear-elastic stresses be high ("stress-intensity factor"), but also the local inelastic stress must exceed a critical value. Thus linear elasticity theory is inadequate for the description of this non-linear phenomenon. Griffith's genius was to recognize that the theory of surface energy is non-linear, and that values for surface energies can be measured independently of cracks. Therefore, energy balances can be used to describe the behavior of cracks, and the description can be tested experimentally.

19.2 Fracture surface energies

The work required to separate a piece of solid into two parts is of more than one kind. If it is done reversibly (which is difficult, but sometimes possible) then it is the work needed to create two new nearly perfect surfaces, and it can be ascribed to intrinsic surface energy. It is, of course, the same thing as the energy of surfaces created in some other way. However, in most cases the separation process is not reversible, so part of the separation work is not reversible, being lost to the generation of heat, material defects, or chemical reactions. Then part of the energy associated with the new surfaces is extrinsic; and the overall surface parameter is said to be the "effective surface energy", or the "fracture surface energy". Furthermore, since the losses can be either small or large, the fracture surface energy can be very much larger than the intrinsic surface energy, sometimes a factor of 1000 (or more) larger.

The intrinsic surface energy is a well defined material property, not so for the fracture surface energy. The latter depends on more than the material itself. It depends on the geometry of a particular specimen, its internal structure, the state of stress (i.e., the components of the stress tensor), the distribution of external tractions, the temperature, the rate of crack propagation through the material, and so on.

The geometry includes the "mode" of a crack. There are the three standard modes (Figure 19.2): the usual *breaking* mode (19.2(a)) in which the principal load is perpendicular to the plane of the crack; the *shearing* mode (19.2(b)) in which the principal load is a couple whose axis is parallel to the tip of a crack; and the *tearing* mode (19.2(c)) where the axis of

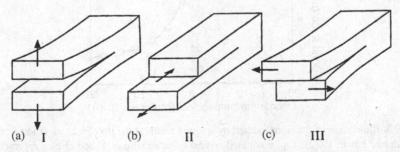

(a)   I   (b)   II   (c)   III

Figure 19.2 Three modes of crack propagation (drawing from Lawn, 1993): (a) mode I, breaking, (b) mode II, shearing, (c) mode III, tearing.

the couple is perpendicular to the axis of the crack tip (Lawn, 1993). The discussion here will concentrate on the breaking mode.

Just as cohesive energies cannot be calculated exactly, neither can surface energies. Approximations must be made. These are different for different types of solids: metal, ionic, covalent, and London types. In all cases they are complicated by the reduction of symmetry at a surface. This leads to the presence of electronic states called Tamm states not found in the interior of the solid. These states have energies somewhat greater than the top of the valence band in semiconductors (so they lie in the gap region), and greater than the Fermi level in metals.

From the perspective of the strength of materials, the extrinsic factors that determine fracture surface energies are perhaps the most important. They determine ductility. And ductility, the ability of a material to withstand deformation without breaking, is critical in determining the reliability of structures, as well as the fabricability of metallic objects by processes such as forging and drawing. However, the extrinsic factors are not independent of the intrinsic factors, so it is important to consider both of them concurrently. This can be understood with the aid of Figure 19.3 which is a schematic view of splitting at the local

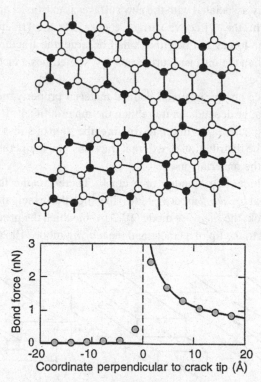

Figure 19.3 Illustration of tractions exerted by strained bonds at the tip of a crack (modified version of Sinclair and Lawn, 1972). Top, computed critical crack in silicon. Filled circles represent atoms in one layer of the structure; open circles represent atoms in the next lower layer. Bottom, forces of bonds resisting breakage apply local traction.

level of a crack tip rather than the overall view of Figure 19.1. The crack converts the rod into two cantilever beams that are "built in" to the block at one end, and loaded by applied forces at the other. The work being done on the rod by the applied forces is resisted by the material near the tip of the crack so as to stop the crack from growing. These forces are clarified in Figure 19.3 which shows a calculated schematic crack in silicon (after Sinclair and Lawn, 1972). The surface forces (stresses) that resist the extension of the crack are shown at the bottom of the figure.

At distances in front of the crack tip that are large compared to the height of the split rod h, there are no net stresses. At large distances behind the crack tip, the stresses are simply those associated with the bending of the two beams. As the crack tip is approached from the right (Figure 19.1) front, the two halves of the rod begin to be displaced outward from one another elastically along the plane of the crack. The resulting strains are largest at the crack "tip" which is distributed over a small region; it is not a point. In the elastic region, the stress increases in proportion to $(1/x)^{1/2}$, where $x < h$, but $x \gg b$ where b is an atomic bond length, or a molecular diameter. Then, if the applied forces F are almost large enough to cause the crack to propagate (at $F = F_{\mathrm{crit}}$), and x becomes comparable with b, the displacements become too large to be described in terms of elasticity theory (Elliott, 1947). In this region the stresses begin to rise in accordance with the chemical bonding forces per unit area, plus the smaller stresses due to the bending of the beams. The bonding stresses follow, to a good approximation, dU/dx, where U is the universal binding energy relation (UBER). So, the part of the stress that is induced by the chemical bond tractions is of the form of the well known "Rydberg function" (Banerjea and Smith, 1988):

$$f(x) = \alpha x (e^{-\beta x}) \tag{19.1}$$

where α and β are constants with α proportional to the elastic modulus, and β approximately equal to $1/b$. Thus $f(x)$ rises initially like $\sin x$ until a maximum is reached and then decays exponentially to zero as the chemical bonding stresses drop to zero, and only the beam stresses remain.

The integral of $U(x)$ from $x = \infty$ to $x = 0$ equals twice the surface energy. This is the work of fracture in the case of a nearly perfect brittle solid (Gilman, 1959). However, if the solid is imperfect so the crack is not flat, then the amount of surface that is generated when the crack tip moves is greater than it would be for a flat crack, and the fracture surface energy is greater than the intrinsic surface energy. This is a relatively small effect, not more than a factor of about two.

19.3 Inelastic effects

If a material responds plastically (or otherwise inelastically) to the stresses induced by $U(x)$, large amounts of work can be done during propagation of the crack tip. This is then absorbed as heat, or internal defects. It can be orders of magnitude greater than the intrinsic surface energy, but it is not independent of the latter.

Some of the inelastic processes that can occur, and that require extra work during crack propagation, include: anelastic relaxations, plastic deformation, twinning, phase transitions, precursor cracking, and microstructure changes such as crazing in polymers. All of these processes are intensified by the stresses derived from $U(x)$. They are minimal at crack tips in layered structures with weak binding between the layers, mica for example. Weak binding means low surface energy, of course. These effects are maximal in tough, nearly isotropic, strain-hardenable solids such as high-strength steels. If the loading rate for cracks in such materials is not too high, the crack-tip zones tend to strengthen as the cracks attempt to advance, so the zones grow and absorb large amounts of energy before they fail. This process depends, however, on substantial levels of intrinsic surface in order to start without premature failure.

It may be seen that high intrinsic surface energy is necessary for a material to have high fracture surface energy (high toughness). Therefore, surface energy is an essential factor in determining the strength of a material. Roughly, if a material with Young's modulus, Y, has a definite yield stress for plastic deformation, σ_y, the fracture surface energy, S_f, is related to the intrinsic surface energy, S_o, by:

$$S_f = (Y/\sigma_y)S_o \tag{19.2}$$

and the smaller the yield stress is relative to the elastic stiffness, the larger is the fracture surface energy. The ratio can of course be quite large, 100 or more.

Like the other factors that determine strength, the intrinsic surface energy has its basis in electronic structure. This will be discussed in Chapter 20, but S_f is an extrinsic property and will not be discussed further.

19.4 Environmental factors

If they are in the environment surrounding a crack, reactive gases, or liquids, may penetrate the crack and react with the atoms at the tip, thereby reducing the effective surface energy. The driving potential for this can be very high. For example, if the free energy of reaction is 1 eV, and the atomic spacing is 2 Å, the equivalent surface energy is 4000 erg/cm^2, and the equivalent driving stress near the tip of the crack is 2×10^{11} dyn/cm^2, or 200 kbar of pressure. This is high indeed.

It is natural to think of the opening of cracks in terms of the "pulling apart" of a material to make two pieces. However, for the operation of environmental effects, the elastic shear strains near the tip of a crack may well be most important. These are large near the tip of a critical crack as may be seen in Figure 19.3. Chemical reactions in, and between, solids are strongly affected by shear strains (Gilman, 2001). This is known as mechanochemistry.

Fractional changes of shear strains affect the chemical stability of a material (i.e., the electronic structure) more than the same fractional change of hydrostatic strain. This is natural since shear affects the shape of a material, as do chemical reactions, so it is consistent with Le Chatelier's Principle. However, the important effects of shear are often overlooked. They are rarely, if ever, discussed in textbooks.

At the tip of a critical crack, the shear strains are largest in the non-linear part of the region. In mode III, the tearing mode, shears are in fact the only large deformations. In mode II, shears are the dominant strains. Only in mode I are both shears and hydrostatic deformations present in approximately equal amounts.

For silicon, Figure 19.3 shows a shear angle as large as 38°. This might well change the electronic structure enough to metallize it, and is certainly enough to destabilize it. It is well known that large shear strains cause silicon to become highly chemically reactive. This is why chemical dissolution occurs preferentially at dislocation cores, forming etch pits.

References

Banerjea, A. and Smith, J.R. (1988). Origins of the universal binding energy relation, *Phys. Rev. B*, **37**, 6632.

Elliott, H.A. (1947). *Proc. Phys. Soc. London B*, **59**, 208.

Gillis, P.P. and Gilman, J.J. (1964). Double-cantilever cleavage node of crack propagation, *J. Appl. Phys.*, **35** (3), 647.

Gilman, J.J. (1959). Cleavage, ductility, and tenacity in crystals. In *Fracture*, ed. B.L. Averbach, D.K. Felbeck, G.T. Hahn, and D.A. Thomas, Chapter 11, p. 193. New York: MIT Technology Press and Wiley.

Gilman, J.J. (2001). Mechanochemical mechanisms in stress corrosion. In *Chemistry and Electrochemistry of Stress Corrosion Cracking: A Symposium Honoring the Contributions of R.W. Staehle*, ed. R.H. Jones, p. 3. Warrendale, PA: The Minerals, Metals and Materials Society.

Griffith, A.A. (1920). The phenomena of rupture and flow in solids, *Philos. Trans. R. Soc. London, Ser. A*, **221**, 163.

Lawn, B. (1993). *Fracture of Brittle Solids*, 2nd edn., p. 24. New York: Cambridge University Press.

Sinclair, J.E. and Lawn, B.R. (1972). *Proc. R. Soc. London, Ser. A*, **329**, 89.

20

Surface and interfacial energies

20.1 Introduction

In the last chapter the intimate connection between fracture and surface energy was discussed for both the ideal and non-ideal cases. The discussion indicated the primary role played by the intrinsic surface energy, S. Now the connection of S to the electronic properties of materials will be emphasized, including the energies of interfaces. Recognizing that all such theories are approximate, only the most simple analytic theories will be presented with brief mention of detailed computer-based but approximate theories. These are sometimes said to be "first principles theories", but they are no more based on "first principles" than other approximate theories based on quantum mechanics. The Heisenberg Theorem ensures this, by limiting the basis of any theory. Also, as this book has mentioned previously, few experimental measurements are accurate enough to justify the application of a high precision theory.

Of general interest is the pattern of surface energies throughout the Periodic Table of the Elements (Figure 20.1). Most of the data are for liquids and are from Murr (1975) and Allen (1972). An obvious exception is the value for diamond which comes from cleavage experiments. Also note that, in the solid state, anisotropic substances like Se have more than one surface energy; one of them is much larger than the value given in the figure, and the other is about the same as the given value. The gaps (e.g., atomic numbers 7–10) are simply places where there are no measured surface energies.

The patterns of Figure 20.1 are similar to those of Figure 12.1 for the bulk moduli. Both show the same series of peaks, the tallest being for covalent carbon. Minimal values occur for the monovalent metals. The usual anomaly occurs for Mn. For the isovalent Groups I through IV, lines can be drawn for which the surface energies are proportional to the functions $1/N^m$ where N is the atomic number and m is a coefficient that depends on the group. This suggests that the surface energy is determined principally by the valence electron density. The next figure (Figure 20.2) testifies to this.

The valence electron numbers for the groups of Figure 20.2 are taken to be the chemical ones: 1, 2, 3, and 4 respectively. Through each set of points for a given group, a straight line can be drawn, and these cluster around the dashed line in the figure. This is consistent with the hyperbolic correlations of the points for each group in Figure 20.1.

254

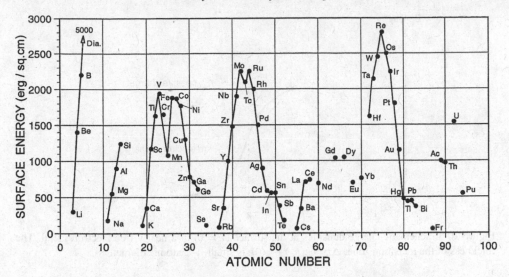

Figure 20.1 Distribution of surface energies in the Periodic Table of the Elements.

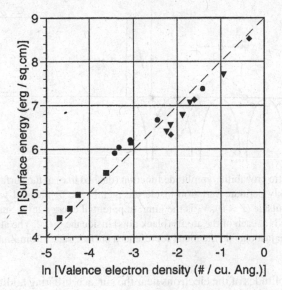

Figure 20.2 Showing, for the sp-bonded elements, the approximately linear dependence of the surface energies on their valence electron densities. The valences used are the chemical valences (1, 2, 3, and 4) for each elemental group (■ Group I, ● Group II, ▼ Group III, ♦ Group IV). The atomic volumes are from Kittel (1996), and the surface energies from Murr (1975).

20.2 Surface states

For a simple metal with a flat free surface, the abrupt stoppage of the array of positive ions creates a step in the positive potential that is seen by the electrons (Figure 20.3). This modifies

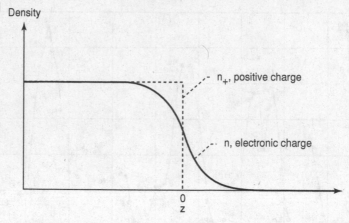

Figure 20.3 Schematic charge densities at the surface ($z = 0$) of a nearly free electron metal. The thickness of the transition zone, $\Delta z = z(n_o) - z(0)$, is small, 1–2 atomic diameters.

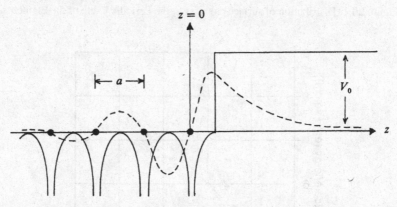

Figure 20.4 Schematic probability amplitude function (dashed line) at the surface of a simple metal. It consists of a weakly periodic variation inside the metal ($z < 0$) blended with an exponentially declining variation outside ($z > 0$). V_o is the jump in potential energy at the surface which is joined onto the potential wells at each of the atoms (black dots) inside the metal. The atom spacing is a. The peak in the amplitude just outside the metal represents a surface state (from Zangwill, 1988).

the probability amplitudes of the electrons near the surface, creating additional energy states called "Tamm states" (Davison and Steslicka, 1992). Near a surface, the frequency of the probability amplitude increases while the magnitude diminishes; thus states with energies higher than those in the interior appear (Figure 20.4). These localized states lie within the gap above the valence band in semiconductors, and above the Fermi energy in a metal. Outside the surface they decay exponentially with distance from the surface. The surface states are important in determining the chemical behavior of a surface. Therefore, they play an important role in environmental effects on fracture, particularly oxidation effects. Discussion of this surface reactivity is beyond the present scope. Interested readers may refer to Zangwill (1988) or to Cohen (1996).

20.3 Surface energies

20.3.1 Ionic crystals

Simple compounds such as the alkali halides have anisotropic surface energies calculated assuming the crystals to be arrays of ions. These are univalent for the alkali halides, but they may also be of higher valence (e.g., divalent for MgO, and mixed trivalent and divalent for Al_2O_3). Since there are alternating positive and negative ions at the surfaces of these crystals, there is a surface Madelung constant, yielding a net electrostatic attraction. In addition, there are Pauli repulsions of the ion cores of the form $+A/\delta^n$, where A and n are constants and δ is the minimum distance between ions.

For example, this approach yielded some of the earliest theoretical estimates of the energies of the principal faces of rocksalt (Born and Stern, 1919):

$$S_{100} = 0.117(q^2/\delta^3) \qquad \text{and} \qquad S_{110} = 0.315(q^2/\delta^3) \qquad (20.1)$$

where q is the electron charge (cgs units), and δ is the interionic distance.

The low energy surface planes (100) are cleavage planes in this case. Figure 20.5 is a polar diagram showing how the surface energy varies with the orientation of the surface (after Yamada, 1924). This two-dimensional plot cannot be extended to three dimensions because separation of a rocksalt crystal on a (111) plane would generate one face with all Na ions, and the other face with all Cl ions. The energies of such faces would be unacceptably large.

20.3.2 Covalent crystals

The surface energies of covalent crystals can be estimated simply by calculating the energies per bond, plus the density of bonds crossing the surface of interest. Then the surface energy is the product of these two quantities divided by two (since there are two surfaces). For homopolar crystals (diamond structure), the lowest energy (cleavage) planes are the

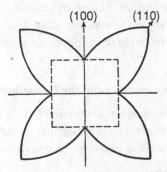

Figure 20.5 Variation of surface energy with orientation of the surface plane for crystals with the rocksalt (NaCl) structure. The polar angle is the angle between the vector normal to the surface and the [100] direction. The minima for the (100) planes are absolute minima, consistent with these being the cleavage planes for rocksalt.

{111} planes because they have the lowest bond densities (number of bonds per unit area). This is not the case for heteropolar crystals (zinc blende structure), because separation of {111} planes generates one face with one of the two constituents covering it, and the other constituent covering the other face. In this case, the planes covered on both sides with alternating species are the {110} planes, and these are the cleavage planes.

20.3.3 Simple metals

For a simple metal, at a free surface, the valence electron density decreases abruptly (Figure 20.3). The decrease occurs over a distance of between one and two atomic diameters. Thus, except for the outermost layer of atoms, the perturbation of the cohesion is not large, and approximate calculations yield good values for the intrinsic surface energies. Three methods will be considered: first, an estimate based on the Heisenberg Principle; second, an estimate based on the elastic properties; and third, a method based on surface plasmon energies. Although these methods are approximate, they are not significantly less accurate than more complex methods based on numerical solutions of the Schrödinger equation (Smith, 1969; also Lang and Kohn, 1970). Also, it should be noted that the experimental values of surface energies are not well known, so there is little justification for precise theoretical calculations when there are few precise experimental values for comparison.

As in the case of the elastic moduli, surface energies are determined primarily by valence electron densities. Therefore, if measured atomic volumes are used, and valences known from chemical behavior, surface energies follow quite directly.

As stated before, the crystalline pattern of discrete positive ions is replaced by a uniform positive electrostatic potential that ends abruptly at the surface. This potential neutralizes the valence electrons (this is a one-electron model) inside the "crystal", where their density is uniformly the number of valence electrons per atom. At the surface where the positive potential ends abruptly, the electrons spill out slightly, creating a thin dipole layer (Figure 20.3). The excess electrons on the outside are balanced by a small deficiency of electron density just inside the surface. The overall thickness of the region where the electron density is decreasing from one to zero per atom is Δ. This equals about an atomic diameter, and is the standard deviation of the positions of the surface electrons.

20.4 Surface energy from the Heisenberg Principle

In order to calculate the surface energy an expression $U(r)$ is needed, relating energy U and atomic size r. The most simple expression is one that we have used before for the energy of an equivalent atom with a uniform electronic charge distribution:

$$U(r) = -\alpha/r + \beta/r^2 \tag{20.2}$$

where α and β are coefficients. The first term is an electrostatic attraction, and the second term is the Schrödinger repulsive pressure. Setting the derivative of this equal to zero, and solving, the equilibrium value of r is found, $r_0 = 2\beta/\alpha$. Substituting this into Equation (20.2) yields the energy at $r = r_0$:

$$U(r_0) = \alpha/2r_0 \tag{20.3}$$

The density of atoms in a square planar array of atoms (b.c.c. crystal structure) is $N = 1/4(r_0)^2$. So the energy of a plane of atoms is:

$$U_p = NU(r_0) = \alpha/8(r_0)^3 = 9q^2/80(r_0)^3 \tag{20.4}$$

When a crystal is cleaved, such a plane becomes divided into two planes, or surfaces. Therefore, since the energy before and after the process must be conserved ($T = 0$ K, so the entropy change is zero) the surface energy S is:

$$S = 9q^2/160(r_0)^3 = 0.056(q^2/r_0^3) \tag{20.5}$$

Taking sodium as an example, $r_0 = 3.659 \times 10^{-8}$ cm, and the electron charge is $q = 4.8 \times 10^{-10}$ esu, so $S_{th} = (9 \times 23.04 \times 10^{-20})/(160 \times 49 \times 10^{-24}) = 264$ erg/cm^2. This compares very well with the experimental value, $S_{exp} = 240$ erg/cm^2. The difference is 10%.

For the full set of alkali metals (valence 1), the dependence of the surface energy on the inverse cube of the atomic spacing is demonstrated in Figure 20.6. The close dependence on the inverse cube of the atomic radius demonstrates that the surface energies are indeed determined by the valence electron densities. This is reinforced by the data of Figure 20.2.

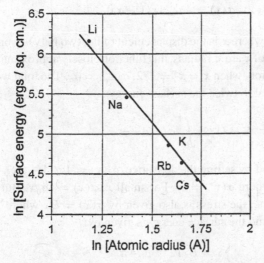

Figure 20.6 Experimental surface energies for the alkali metals as a function of their atomic radii. Ln–ln plot with a regression line of slope −3. Showing that the surface energy is proportional to the valence electron density, consistent with the theory.

Applying this result to fracture behavior, it may be seen that the fundamental basis for the resistance of solids to brittle fracture lies in just a few principles: Coulomb's Law of electrostatics, Heisenberg's Theorem, and Pauli's Principle. In the case of semi-brittle solids, that is, cracks moving through real engineering materials, considerably more energy is absorbed than the intrinsic surface energy (100–1000 times as much). The additional energy is a function of the underlying intrinsic energy absorption which stimulates various extrinsic absorption mechanisms. The most important of the latter is plastic deformation. Next in order of importance is the chemical reactions that occur between the nascent surfaces of fresh cracks and the environment to which they are exposed. These extrinsic mechanisms are driven by the stresses that resist the formation of new surfaces. This is why it is important to understand the basis of intrinsic surface energies.

20.5 Surface energy from elastic stiffness

The experimental input for the previous method was minimal. It only used the atomic radius, plus some knowledge of chemical valence. Now, more experimental information will be used, namely, elastic stiffnesses plus some straightforward logic.

At low temperatures, if a block of solid is reversibly cleaved into two pieces, the work of cleavage is partitioned equally between the two surfaces so half of it equals the surface energy. As the material of the block begins to separate, linear elastic forces resist the separation. Then the resistive force increases less rapidly and eventually reaches a maximum where it begins to decrease at first slowly and then rapidly until it approaches zero asymptotically. A simple algebraic function:

$$\sigma(x) = 4\sigma_{max}[x(1-x)] \tag{20.6}$$

approximates this behavior where x is the displacement of the two halves from the equilibrium spacing d_0. If the units of x are π radians, this function closely approximates $\sigma_{max} \sin x$ for $[0 \le x \le 1]$. By inspection, when $x = x^\star = 1/2, \sigma(x^\star) = \sigma_0$. The total work done as x increases from 0 to 1 is equal to twice the surface energy:

$$S = (1/2) \int_0^1 \sigma(x)\,dx = \sigma_0/3 \tag{20.7}$$

The maximum force is found by setting the derivative, $\sigma'(x) = 4\sigma_0(1 - 2x)$, equal to zero and solving for $x^\star = 1/2$ where $\sigma(x^\star) = \sigma_0$. For small $x, \sigma(x) = 4\sigma_0 x$. But since the response, for small x, is elastic, the stress is also given by $\sigma(x) = Ex$ where E is Young's modulus. Then $\sigma_0 = E/4$, and the surface energy is given by:

$$S = E/12 \approx E/4\pi \tag{20.8}$$

This simple expression gives surprisingly good values for S. It can also be used for anisotropic crystals if $E = E(hkl)$ where the latter is the orientation dependent modulus.

20.6 Surface energy from plasmon theory

An entirely new and powerful approach to surface energy theory was developed about 30 years ago by Schmit and Lucas (1973). This theory has been somewhat controversial (Kohn, 1973; and rebuttal, Phillips, 1975), but it is consistent with a large amount of data, and with earlier theories. Plasmon energies depend on valence electron densities, just as do elastic stiffnesses and surface energies. An important advantage of this theory is that it provides a simple way to obtain interface (adhesion) energies.

Plasmons are fundamental excitations in all solids (Pines, 1963). They were discovered experimentally by R.W. Wood (1933) who found that the alkali metals become transparent to ultraviolet light at sharp "transparency edges". These critical wavelengths were soon given a simple theoretical interpretation, that still stands, by Zener (1933). He pointed out that the transparency edges are associated with collective oscillations of the plasmas (valence electrons – positive ions) in metals. These oscillations have characteristic quantized energies $\hbar\omega_p$ (where $\hbar$ is Planck's constant divided by 2π, and ω_p is the plasma frequency). The quantum states of the oscillators are given by $(n + 1/2)\hbar\omega_p$ with $n = 0, 1, 2, 3, \ldots$, so the zero-point energy is $(\hbar\omega_p/2)$. Thus oscillations of this energy ($n = 0$) are always present. For light with energy greater than $\hbar\omega_p$, states with $n = 1$, and greater depending on the energy, can be excited. These can then reradiate the light so it can propagate through the metal. For less energetic light of frequency $\omega < \omega_p$, the dielectric constant becomes negative (and the refractive index imaginary) so incident light is reflected. The wavelength of the light is long compared with the wavelength of a valence electron, so the displacement associated with the oscillator is that of a collective group of electrons.

According to Zener's theory, the following expression gives the plasmon frequency ω_p (cgs units):

$$\omega_p^2 = 4\pi N q^2/m \tag{20.9}$$

where N is the electrons per cm^3, q is the electron charge, and m is the electron mass. The corresponding wavelength λ_T of the transparency edge is given by (c is the speed of light):

$$\lambda_T^2 = \pi(m/N)(c/q)^2 \tag{20.10}$$

For sodium, using the free-electron charge and mass, this agrees remarkably well with the measured value: λ_T(theory) = 2090 Å, λ_T(experiment) = 2100 Å.

Plasmons can be conveniently observed in electron microscopes equipped with electron energy spectrometers. The beam of known energy is passed through a thin specimen, and the energy of the transmitted beam is measured. At the plasmon energy there is a peak in the amount of energy that is lost. This is called EELS, or electron energy loss spectroscopy (Egerton, 1996).

In addition to bulk plasmons, there are surface plasmons. If a block of metal is cleaved into two pieces, and the distance between the two pieces is z, then surface plasmons ω_s appear on the fresh surfaces. Their electric fields decay rapidly in the direction z normal to the surface, and their frequencies are given by:

$$[\omega_s(z)]^2 = (\omega_p^2/2)(1 - e^{-kz}) \tag{20.11}$$

where k is the wave number of a valence electron, $\approx 1/$atomic diameter. Thus, the surface plasmons are fully formed for very small separation distances. After they have been formed, they will attract one another by exchanging photons. This is the mechanism of the van der Waals attraction and diminishes in proportion to d^{-6}.

The interaction of the plasmons between closely spaced cleaved surfaces may be important in determining the ductilities of practical materials. It is consistent with the observation that materials with high polarizabilities (low shear moduli) tend to be more ductile than those with low polarizabilities.

The Schmit–Lucas theory considers what happens to the plasmon modes when a block of solid is split into two parts. For each part, one bulk mode is lost, while one surface mode is created at the cleavage surface. The energy of the system thereby increases according to Equation (20.11). Both are zero-point modes of the oscillators. This energy shift provides a good estimate of the surface energies of more than 50 metals (Schmit and Lucas, 1973).

20.7 Interfacial energies from plasmon theory

A useful feature of the plasmon theory is that it gives a very simple means for estimating the energies of interfaces. Since these lie between unlike substances, they are considerably more complicated than homogeneous interfaces. For a pair of substances, A and B, the individual plasmon energies can be measured by means of EELS. Then, from straightforward electromagnetic theory, it is known (and verified experimentally) that the energies of surface plasmons, u_s, are related to the energies of bulk plasmons, u_b, by $u_s^2 = u_b^2/2$ (Kittel, 1996). Also, the energy of an interfacial plasmon, u_{AB}, is simply the rms of the surface plasmon energies, u_{sA} and u_{sB}, of the two solids. That is:

$$u_{AB}^2 = (u_{sA}^2 + u_{sB}^2)/2 \tag{20.12}$$

Therefore, since surface energies are proportional to the 3/2 powers of plasmon energies (Gilman, 1999), the interfacial surface energy, S_{AB}, is given in terms of the individual surface energies, S_A and S_B, by:

$$2S_{AB}^{4/3} = (S_A^{4/3} + S_B^{4/3}) \tag{20.13}$$

20.8 Long-range attraction of cleavage faces

The collective polarizations that form plasmons are analogous with the polarizations of molecules. Therefore, polarizability governs both of them; and both of them are related to shear stiffness. A consequence of the dipole–dipole forces that act between the polarization oscillators is that the decline in the force that acts between the two separating surfaces at a crack tip is less rapid when the polarizability of the solid is large. Therefore, when a crack is propagating at a given velocity there is more time for plastic deformation if the polarizability of the crystal is large. In effect, the "zone" of separation at the tip of a crack is longer when the polarizability is larger. This, in turn, lengthens the region in which plastic deformation (and other inelastic processes) may occur, thereby toughening the material.

At distances which are large compared with the atomic diameters, induced atomic dipole–dipole forces (London forces) become more important than the monopolar forces within the atoms. The latter decline exponentially with distance d, while the former decline as $1/d^6$. So, although they are much smaller, the dipole forces eventually prevail. If d is the atomic diameter, the crossover occurs between $9d$ and $10d$.

At intermediate distances, the interactions between the fluctuating plasmon dipoles become dominant. These interactions also provide an alternative means for describing the macroscopic forces between solids caused by charge fluctuations. These were first described theoretically by Lifshitz (1956), and measured experimentally by Derjaguin, Churaev, and Muller (1987). As mentioned above, surface plasmons also provide an alternative theory of surface energy (Schmit and Lucas, 1973). Furthermore, Gerlach (1971) has shown that plasmon interactions lead to the same result as that of Lifshitz for the macroscopic forces. Finally, Chen (1993) has shown that the forces at intermediate distances can be described in terms of the tunneling of electrons back and forth across the gap between two solids. A net result of these various theories is that they indicate the considerable importance of polarizability for fracture behavior.

The mutual attraction of two polarizable particles, a and b, was discussed in a previous chapter, and will be reviewed here (this is the Drude theory, after Rigby *et al.*, 1986). The particles are electrostatically neutral, and the centers of positive and negative charge in them have zero separation, z_a, z_b, on average. However, the separations in each of them can fluctuate, so each particle forms a harmonic oscillator with a fluctuating dipole moment. The oscillators have spring constants, k, and masses, m, equal to the electron mass for valence 1. It is assumed that both particles are of the same kind. Let the orientations of the dipole moments be colinear since this yields the strongest interaction (Figure 20.7). The distance between the particles is d.

When $d \gg z_a$ or $d \gg z_b$, so the oscillators are independent, their frequencies are $v = (1/2\pi)(k/m)^{1/2}$, but as d decreases (remaining larger than the z values) the oscillators begin

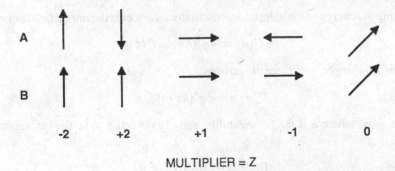

MULTIPLIER = Z

Figure 20.7 Schematic diagram of a linear pair of dipoles.

to interact and two oscillation frequencies develop as energy is exchanged (via photons) between the two oscillators.

The oscillators under consideration are quantized, so they have zero-point energies (i.e., they oscillate at all temperatures). The Schrödinger equation for one of these oscillators is:

$$\frac{1}{m}\frac{\partial^2 \psi}{\partial z_a^2} + \frac{8\pi^2}{h^2}\left(E_a - \frac{kz_a^2}{2}\right)\psi = 0 \tag{20.14}$$

The quantized energy values given by this are $E_a = (n_a + 1/2)h\nu$. Similarly, for the other oscillator, $E_b = (n_b + 1/2)h\nu$. And, for the lowest energy of the system, when $n_a = n_b = 0$, $E(d \gg z) = 2(h\nu/2) = h\nu$.

When the oscillators are close enough to interact significantly, if q is the electron charge, the interaction energy for the system of two oscillators is $(2q^2 z_a z_b)/d^3$ (cgs units), and the Schrödinger equation becomes:

$$\frac{1}{m}(\nabla_a^2 + \nabla_b^2)\psi + \left[E - \frac{k}{2}(z_a^2 + z_b^2) - \frac{2z_a z_b q^2}{d^3}\right]\psi = 0 \tag{20.15}$$

The interaction term changes the frequencies, and therefore the energies. The new spring constants are:

$$k_1 = k - 2q^2/d^3$$
$$k_2 = k + 2q^2/d^3$$

and the new frequencies:

$$\nu_1 = \nu[1 - 2q^2 d^3/k]^{1/2}$$
$$\nu_2 = \nu[1 + 2q^2 d^3/k]^{1/2}$$

The lowest energy for the system is now the sum of the new zero-point energies:

$$E(d) = (h/2)(\nu_1 + \nu_2)$$

Expressing the frequencies as binomial series, and keeping only the first term:

$$E(d) = h\nu[1 - q^4/2d^6 k^2]$$

Subtracting the energy of the independent oscillator case yields the interaction energy, $U(d)$:

$$U(d) = -q^4(h\nu)/2d^6 k^2 \tag{20.16}$$

and if this is averaged over three dimensions:

$$U(d) = -3q^4(h\nu)/4d^6 k^2$$

But $q^2/k = \alpha$, where α is the polarizability, and $(h\nu)$ is taken to be the ionization energy U_I, so:

$$U(d) = -(3/4)(\alpha^2/d^6)U_I \tag{20.17}$$

which indicates an interaction energy that increases strongly with α, and decreases very strongly with distance d.

So much for atomic dipole interactions. Of more interest here are the interactions that occur between the large aggregates of dipoles in pieces of solids.

Consider two solid slabs of material with polarizability α placed with a gap d between them. For simplicity they both have the same values of α. Each slab contains a large number N of fluctuating (induced) dipoles. The individual dipoles interact as pairs with attractive energies U_p:

$$U_p = -3\alpha^2 U_I/4d^6$$

For the two slabs, N_a dipoles in one slab interact with N_b dipoles in the other slab. Therefore, the interaction energy of the slabs is:

$$U_s = \int_a \int_b N_a N_b \left(\frac{3\alpha^2 U_I}{4d^6} \right) dV_a \, dV_b$$

$$= -\pi N_a N_b \alpha^2 U_I/16d^2 \tag{20.18}$$

which indicates that the interaction declines much less rapidly with distance than for individual atomic particles, and still depends strongly on the polarizability.

20.9 Importance of polarizability

The important role of polarizability in the fracture process is one reason why computed simulations of fracture processes are not effective. They are usually based on the interactions of individual particles. This neglects the many particle interactions associated with polarizability.

A good illustration of the importance of polarizability is provided by the set of halide crystals. Through the series of 20 alkali halides, from the least polarizable (LiF) to the most polarizable (CsI) the cleavability clearly declines. Also, highly polarizable halides such as the silver salts (AgCl, AgBr, and AgI), and the thallium halides (TlCl and TlBr) cannot be cleaved at all.

In the case of metals, the correlation between G/B and brittleness is another consequence of polarizability.

Also, the toughness of polymeric systems like polycarbonate (Lexan) is related to its high polarizability. This is also true of tough adhesives.

All of these behaviors indicate that polarizability has an important effect on ductility. As mentioned at the beginning of this section, it seems likely that it does this by lengthening the "process zones" at the tips of propagating cracks. This zone is the region where the concentrated stresses are high, and plastic deformation occurs. The latter is a time dependent process so the longer the zone the more time there is for plastic deformation at a given crack velocity.

Dislocation mobility is involved, along with the effect of polarizability in determining the plastic deformation at a crack tip. If it is low, plastic flow is slow and the length of the process zone does not matter, so high polarizability has little effect. However, as was shown in Chapter 13, the shear modulus depends on the reciprocal polarizability, so the two effects tend to reinforce one another.

References

Allen, B.C. (1972). In *Liquid Metals*, ed. S.Z. Beer. New York: Marcel Dekker.

Born, M. and Stern, O. (1919). *Sitzungsber. Preuss. Akad. Wiss.*, (January–June), 901.

Chen, C.J. (1993). *Introduction to Scanning Tunneling Microscopy*. Oxford: Oxford University Press.

Cohen, M.H. (1996). Strengthening the foundations of chemical reactivity theory. *Topics in Current Chemistry*, Volume 183, p. 143. Berlin: Springer-Verlag.

Davison, S.G. and Stęslicka, M. (1992). *Basic Theory of Surface States*. Oxford: Clarendon Press.

Derjaguin, B.V., Churaev, N.V., and Muller, V.M. (1987). *Surface Forces*, trans. V.I. Kisin and J.A. Kirchener. New York: Consultants Bureau.

Egerton, R.F. (1996). *Electron Energy Loss Spectrometry in the Electron Microscope*, 2nd edn. New York: Plenum Press.

Gerlach, E. (1971). Equivalence of van der Waals forces between solids and the surface–plasmon interaction, *Phys. Rev. B*, **4**, 393.

Gilman J.J. (1999). Plasmons at shock fronts, *Philos. Mag. B*, **79** (4), 643.

Kittel, C. (1996). *Introduction to Solid State Physics*, p. 302. New York: Wiley.

Kohn, W. (1973). *Solid State Commun.*, **13**, 323.

Lang, N.D. and Kohn, W. (1970). Theory of metal surfaces; charge density and surface energy, *Phys. Rev. B*, **1**, 4555.

Lifshitz, E.M. (1956). *Sov. Phys. JETP*, **2**, 73.

Murr, L.E. (1975). *Interfacial Phenomena in Metals and Alloys*. Reading, MA: Addison-Wesley.

Phillips, J.C. (1975). Surface energy of metals, *Comments Solid State Phys.*, **6**, 91.

Pines, D. (1963). *Elementary Excitations in Solids*. New York: WA Benjamin.

Rigby, R., Smith, E.B., Wakeham, W.A., and Maitland, G.C. (1986). *The Forces between Molecules*. Oxford: Oxford Science Publications.

Schmit, J. and Lucas, A.A. (1973). Continuum surface waves and surface energy, *Collective Phenom.*, **1**, 127.

Smith, J.R. (1969). Self-consistent many electron theory of electron work functions and surface potential characteristics for selected metals, *Phys. Rev.*, **181**, 522.

Wood, R.W. (1933). *Phys. Rev.*, **44**, 353.

Yamada, M. (1924). *Phys. Z.*, **25**, 52.

Zangwill, A. (1988). *Physics at Surfaces*, p. 64. New York: Cambridge University Press.

Zener, C. (1933). *Nature*, **132**, 968.

21

Fracturing rates

21.1 Introduction

Most fracture is athermal, either because it occurs at low temperatures, or because it occurs too fast for thermal activation to be effective. Thus it must be directly activated by applied stresses. This can occur via quantum tunneling when the chemical bonding resides in localized (covalent) bonds. Then applied stresses can cause the bonding electrons to become delocalized (anti-bonded) through quantum tunneling. That is, the bonds become broken. The process is related to the Zener tunneling process that accounts for dielectric breakdown in semiconductors. Under a driving force, bonding electrons tunnel at constant energy from their bonding states into anti-bonding states. They pass through the forbidden gap in the bonding energy spectrum.

21.2 Thermal activation

At elevated temperatures, the process known as stress-rupture occurs. It is a result of thermally activated vacancy motion. It will not be discussed in detail here. In order for atoms to move from one atomic site in a crystal to another, the temperature must be relatively high, usually above the Debye temperature where the vibrations of individual atoms are excited, and vacancies can be thermally generated. The relatively large masses of atoms prevent much activity at lower temperatures.

The mass of an electron is $1/2000$ times smaller than the mass of the smallest atomic nucleus, the proton. It is about $1/20\,000$ times lighter than a carbon atom, and $1/200\,000$ times lighter than a molybdenum atom. Even at $T = 0$ K, when an electron is localized to atomic dimensions, it has a relatively large amount of kinetic energy. A typical Debye temperature might be 100 K where the energy of an atomic vibrational mode is $kT/2 = 4.3 \times 10^{-3}$ eV. However, at the same temperature, an electron confined to about 5 Å would have a kinetic energy of about $2 \times 10^{+3}$ eV, or about $500\,000$ times larger. Needless to say, as a result of their small masses and high energies, electrons are considerably more nimble than atoms.

At high temperatures various thermally activated processes can lead to fracture. By "high" is meant temperatures at which local interatomic vibrations become excited. For most crystals, most of these modes are activated when the temperature reaches the Debye temperature. However, the fraction of atomic modes activated drops off rapidly for temperatures below this. Thus, for a temperature about 14% below the Debye temperature only about half of the atomic modes are activated. These are the important modes, of course, for the fracture of covalent bonds, for local chemical reactions, and the like. Longer wavelength modes have considerably less average intensity at any given atomic site.

For polymeric solids, the temperature of importance is that at which intra-chain vibrations become active. Call this temperature T_E. For the C–C bonds of polyethylene this temperature is about 1300 K. At room temperature (300 K) only about 3% of the longitudinal vibrational modes are excited to an energy of $kT/2 = 0.013$ eV. Compared with the bond dissociation energy of 3.6 eV, this is only 0.4%. Fluctuations of the thermal energy of this magnitude, according to the Boltzmann distribution function, occur infrequently. At room temperature, the probability is $[\exp -(40 \times 3.6)] = 10^{-63}$. If the atomic vibrational frequency is about $5 \times 10^{+12}$, the rate is about 10^{-51} per second. So, even if there are 10^{+22} bond sites/cm^3, the rate is only 10^{-33} cm^{-3} s^{-1}, or once per cm^3 per $3 \times 10^{+25}$ years. This is very much longer than the estimated age of the universe.

Under an applied stress, the strain energy in a bond of a polyethylene chain becomes equal to the thermal energy per bond when the strain becomes about $(kT/Bb^3)^{1/2} \approx 0.059$, say 6% (the corresponding stress is about 10 GPa) where kT is the thermal energy, B is the bulk modulus, and b^3 is an atomic volume. Therefore, in many circumstances leading to fracture, the strain energy per atom is comparable with, or exceeds, the thermal energy. Then, as will be shown, since the strain raises the electrostatic chemical potential, electron tunneling from a bonding state to an anti-bonding state becomes a favorable process.

Unlike most atoms (H is the exception), electrons can pass through (rather than over) energy barriers. A simple example is what happens at the interface between the two metal contacts in an ordinary household electrical switch. The contacts might be made of brass. Then each of them has a thin layer (about 1000 Å thick) of zinc oxide on its surface. But zinc oxide is an insulator. So how does electricity pass through the interface? It tunnels through. For an electron on one side of the insulating barrier, its probability amplitude on the other side is small but not zero. Therefore, there is a finite probability for finding the electron on the other side. This probability is proportional to the square of the ratio of the amplitude on the "other" side to the amplitude on the first side. This ratio increases exponentially with the voltage difference across the barrier. This, plus the high frequency with which the electron attempts to penetrate the barrier (about 3×10^{16} s^{-1} for 120 V applied across a 100 Å oxide barrier), yields the observed current through the contact barrier.

A similar process, called Zener tunneling, can occur at the bonds between atoms in a molecule (Zener, 1934). In this case, the barrier is the LUMO–HOMO energy gap in which there are only imaginary electron states, and therefore no electrons. Then, in order for an electron to cross the gap it must gain an amount of energy equal to that of the gap (perhaps from a photon). But if a potential is applied (either an electric field, or its equivalent, a stress) the LUMO level decreases, and at a high enough applied potential, its level becomes equal to the HOMO level. When this happens a bonding electron at the HOMO level can tunnel across the gap into the anti-bonding level of the same energy on the other side. Even though the probability amplitude of the bonding electron is small on the LUMO side, when it is combined with the large attempt frequency, the overall tunneling rate becomes significant. When a bonding electron tunnels to an anti-bonding state, the bond that it represents is broken, and the atoms that it connected become surface atoms; or in some cases, they become free-radicals (Gilman and Tong, 1971).

The theory of quantum tunneling leads to a characteristic rate law in which the tunneling probability is a function of the virtual work of fracture instead of being a function of the thermal energy as in thermal reaction-rate theory. And, the attempt frequency is determined by the density of Zener–Bloch states rather than being determined by thermal vibration frequencies (Ridley, 1999). Through these parameters experimental results can be compared with the theory. This will be done here for a variety of crystals, glasses, and polymers.

In addition to being related to the theory of dielectric breakdown, this theory of fracture is also related to the Fowler–Nordheim theory of the field emission of electrons. The only essential difference is that the bonding electrons are emitted into a vacuum in field emission, whereas they are emitted into electronic surface states during fracture.

The localized fracturing of bonds occurs in a stressed specimen prior to the time when the Griffith criterion for crack propagation is satisfied. Thus, a stressed specimen can contain many broken bonds before the overall specimen breaks. The broken bonds can be detected by the spin resonance of the free-radicals that the broken bonds become.

Evidence for the validity of this approach to low temperature fracture has been supported by the report of Doremus (1994) that the rate equation derived from it agrees with experimental results better than other theories. Also, recent data for Kevlar agree with the theory as shown here.

21.3 Fracture via tunneling

The situation at the tip of a sub-critical crack (a crack shorter than a critical Griffith crack) is shown schematically in Figure 21.1. The crack tip does not move concertedly (as a rigid line). It moves in discrete jumps as bond after bond becomes broken until the crack length reaches the Griffith critical value. In the super-critical regime, the rate of crack growth is limited either by the inertia of the medium, or by viscous drag, so it accelerates until it reaches a velocity of about $v_l/3$ where v_l is the speed of a longitudinal sound wave.

In the sub-critical regime, the net crack velocity, v_c, results from the sum of the velocities of the individual kinks along the crack tip, v_k, and the concentration of kinks (fraction of occupied sites), c_k, so that $v_c = c_k v_k$. The cross-sectional area of each kink is ab where b is the distance between bonds along the tangent of the crack tip, and a is the distance between bonds perpendicular to the crack tip. Then, if the bond length is c, the relaxation volume associated with a unit of motion, b, of the crack kink is abc, and the kink velocity is $v_k = db/dt$.

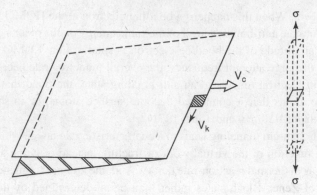

Figure 21.1 Schematic kink on the tip of a crack.

Let the local Cartesian coordinates be $x \| a$, $y \| b$, and $z \| c$. Also, let the principal applied stress be σ_{zz} so the local stress at the kink is $\sigma_c = C\sigma_{zz}$ where C is the stress concentration factor. Then, the strain energy, U_k, of the kink region is approximately:

$$U_k = \left(\sigma_c^2/2Y\right)abc \tag{21.1}$$

where Y is Young's modulus. This is the chemical potential of the kink. The force tending to move the kink in the direction that extends the crack is:

$$\partial U_k/\partial y = (abc/Y)(\partial \sigma_c/\partial y)\sigma_c = f_\sigma \tag{21.2}$$

This is a generalized force. Since the hypothetical fracture mechanism begins with the promotion of a bonding electron into an anti-bonding state, it is equivalent to the electrical force needed to do that. The electrical force, f_e, on an electron of charge q, is Eq, where E is an electric field. Therefore $f_e = f_\sigma$, and:

$$Eq = ac\sigma_c \approx Cb^2\sigma_{zz} \tag{21.3}$$

where b is the bond length. This expression indicates that there is a simple connection between a stress field and an electric field, allowing theoretical results from the quantum mechanics of electrons to be applied to mechanical problems.

When a valence (bonding) electron undergoes a transition to an anti-bonding state, the local potential changes from attractive to repulsive. A simple, and approximately correct, way to represent this is by means of a bonding Morse potential, together with its conjugate anti-bonding potential (Keyes, 1975). One is obtained from the other simply by changing the sign of one of the two terms. This was discussed in Section 10.2.5. From Equations (10.20) and (10.21), at the equilibrium separation, $r = b$, $f = 1$, so $E_b = -D$, and $E_a = +3D$. Then: $\Delta E = 4D$, and the force tending to separate the atoms is $-(dE_a/dr)_{r=b}$. This can be related to the curvature of E_b at $r = b$ (the bulk modulus, B) using relationships given by Keyes (1975). If it is expressed as a pressure by dividing the force by b^2, the answer is that the dissociation pressure, $P_{dissoc} \approx -B$ which is of the order of megabars. Thus, if a bond is put into an anti-bonding state, a very large pressure tends to break it. This is a quantum mechanical effect caused by the large kinetic energy of a nearly free-electron in an anti-bonding state. The pressure is sometimes called the "Schrödinger" pressure.

21.4 Zener tunneling

In order to account for the dielectric breakdown of solids when they are subjected to very large electric fields, Zener (1934) proposed that breakdown avalanches could start with the tunneling of electrons from the valence band where they are immobile to the conduction band where they become mobile. When they have become mobile, the applied field can accelerate them to high enough velocities to excite more electrons through collisions, causing the electron avalanches that constitute breakdown.

Similarly, Gilman and Tong (1971) proposed that this kind of band-to-band tunneling can cause fracture by breaking covalent bonds; the difference in this case is that the stress field is the driving force instead of an electric field.

There are three factors to consider: the electronic attempt frequency ν_e, the tunneling probability p_t, and the acoustic attempt frequency ν_a (which may be much slower than the electronic frequency). In one dimension, the electronic frequency is associated with Zener–Bloch oscillations, Figure 21.2, and is given by (Ridley, 1999)

$$\nu_e = (abc)\sigma_c/h \tag{21.4}$$

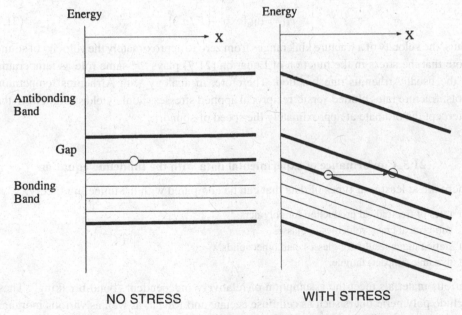

Figure 21.2 Schematic diagram of Zener band-to-band tunneling.

with h Planck's constant, and using Equation (21.3). For a high applied stress, say $Y/8\pi$, and $abc = 8 \times 10^{-24} \text{ Å}^3$, since $h = 6.6 \times 10^{-27}$ erg s, this attempt frequency is about 10^{13} s^{-1}.

The acoustic attempt frequency is determined by the speed of sound, $v_s \approx \sqrt{(C_{11}/\rho)}$, and the bond length, b:

$$v_a = v_s/b \qquad (21.5)$$

which is comparable with, but smaller than, v_e.

Next, the tunneling probability needs to be considered. This depends very sensitively on the height of the barrier to be penetrated, and on the magnitude of the applied stress, but is relatively insensitive to the shape of the barrier (Bell, 1980). Therefore, the shape will not be considered further. The barrier is simply taken to be the energy gap in the bonding energy spectrum (the band gap, or the LUMO–HOMO gap). In the absence of an applied stress the energy diagram is shown on the left in Figure 21.2 where the coordinates are E (energy) and x (reaction path). The effect of an applied stress on the total energy is shown in Figure 21.2 on the right. It may be seen that an electron can now transfer from the top of the valence band to the bottom of the conduction band (or from the HOMO level to the LUMO level) at constant total energy. However, to do this, it must tunnel through the forbidden gap. This is the problem that Zener addressed. Transposing his result from the case of an applied electric field to the case of an applied stress field gives the tunneling probability:

$$p_t = \exp -\{(\pi E_g/h)^2/b\sigma_c\} \qquad (21.6)$$
$$= \exp -(F/\sigma_c)$$

and the expression for the fracture kink velocity, v_{kf}, is:

$$v_{kf} = cv_a[\exp -(F/\sigma_c)] \qquad (21.7)$$

Thus, the velocity of a fracture kink ranges from zero to approximately the velocity of sound. Note that the stress in the function of Equation (21.7) plays the same role as temperature in the usual Arrhenius rate function. Therefore, in analogy with Arrhenius temperature plots, fracture rates plotted versus reciprocal applied stresses should yield straight lines that intercept the ordinate at approximately the speed of sound.

21.5 Conformance of experimental data with the tunneling equation

There are at least four types of data that can be compared with the tunneling theory:

(1) rates of free-radical production for polymers,
(2) sub-critical crack velocities in glasses,
(3) static fatigue results for glasses, and other solids,
(4) dynamic (impact) fatigue.

Various materials meet the assumption of relatively independent "bond-breaking". These include polymeric fibers such as cellulose acetate and Kevlar, as well as various inorganic

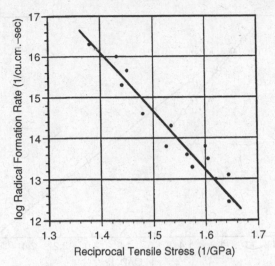

Figure 21.3 Rate of free-radical formation in stressed Capron fibers. The straight line conforms with Equation (21.7).

glasses including fused silica. Some of the data have been reviewed previously by Gilman and Tong (1971), but new data have appeared in the meantime. Doremus (1994) has reported systematic studies of the equations that have been used in the literature by various authors to describe the fracture of glass. He concludes that Equation (21.7) gives the best description of the available data.

When a polymer molecule breaks, the electron-pair bond becomes two free-radicals on the broken ends. These can be detected as a function of time prior to the formation of a critical crack by a spin resonance spectrometer; and this rate depends on the magnitude of the applied stress. Data for this effect, reported by Zhurkov and Tomashevskii (1966) are shown in Figure 21.3 plotted to conform with Equation (21.7). That is, the logarithm of the rate versus the reciprocal stress is plotted. It may be seen that the conformance is quite good over more than four orders of magnitude in the rate.

Cracks that are put into small plates of glass and then subjected to stresses somewhat less than the critical stresses needed to make them propagate rapidly are observed to propagate slowly. Their velocities depend sensitively on the applied stress as expected from Equation (21.7). Data obtained by Wiederhorn (1970) for two examples are given in Figure 21.4. These data conform well to the form of the equation. Note that the velocity data in the figure extend over five orders of magnitude.

Two types of data exist for fatigue tests: static in which constant (nominal) stresses are applied and the time to failure is recorded; and dynamic in which repeated light impacts are applied and the number of impacts (time) to failure is recorded. Figure 21.5 shows impact fatigue data (obtained by Findley and Mintz) for cellulose acetate fibers as reported by Taylor (1947).

Fracture resistance

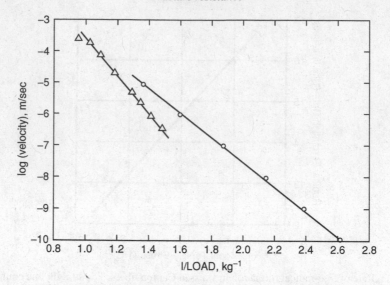

Figure 21.4 Rate of crack propagation in soda-lime glass for two humidity conditions.

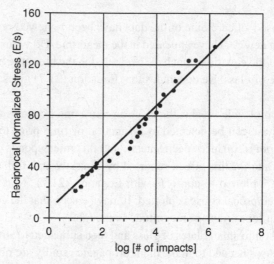

Figure 21.5 Impact fatigue of cellulose acetate fibers. The normalized stress is the applied stress (psi) divided by Young's modulus (1.95×10^5 psi). The logarithm is base 10.

The most extensive data in the literature are those for static fatigue for various glasses, ceramics such as porcelain, and various environmental conditions. A considerable set of these data follow the Preston–Haward law (Glathart and Preston, 1946), and Gilman and Tong (1971) pointed out that rearrangement of this law yields Equation (21.7). More recent data of Pavelchek and Doremus (1976) for soda-lime glass are shown in Figure 21.6. Doremus' book (1994) presents other examples such as fused silica, and borosilicate glass, all of which behave consistently with Equation (21.7).

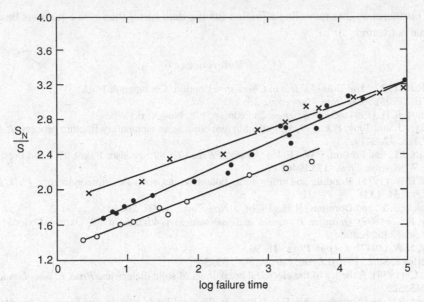

Figure 21.6 Static fatigue data of Burke, Doremus, Hillig, and Turkalo for specimens of soda-lime glass. S_N is the short time breaking stress at a low temperature (78 K). From Doremus (1994).

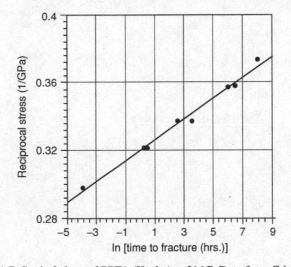

Figure 21.7 Static fatigue of PPTA (Kevlar) at 21 °C. Data from Crist (1995).

Another polymeric material that behaves as expected by the present theory is Kevlar (Crist, 1995). Figure 21.7 illustrates this.

Evidently, at relatively low temperatures, covalently bonded materials fracture as a result of stress-induced quantum-mechanical tunneling of electrons from bonding states into anti-bonding states. The resulting rate equation is consistent with a large variety of experimental data. Furthermore, the tunneling equation fits the data better than rate equations based on thermal activation.

References

Bell, R.P. (1980). *The Tunnel Effect in Chemistry*. London: Chapman & Hall.

Crist, B. (1995). *Annu. Rev. Mater. Sci.*, **25**, 295.

Doremus, R.H. (1994). *Glass Science*, 2nd edn., p. 178. New York: Wiley.

Gilman, J.J. and Tong, H.C. (1971). Quantum tunneling as an elementary fracture process, *J. Appl. Phys.*, **42**, 3479.

Glathart, J.L. and Preston, F.W. (1946). *J. Appl. Phys.*, **17**, 189; see also, Baker, T.C. and Preston, F.W., *J. Appl. Phys.*, **17** (1946) 170.

Keyes, R.W. (1975). Bonding and antibonding potentials in Group-IV semiconductors, *Phys. Rev. Lett.*, **34**, 1334.

Pavelchek, E.K. and Doremus, R.H. (1976). *J. Non-Cryst. Solids*, **20**, 305.

Ridley, R.K. (1999). *Quantum Processes in Semiconductors*, 4th edn., p. 50. Oxford: Oxford Science Publications.

Taylor, N.W. (1947). *J. Appl. Phys.*, **18**, 943.

Wiederhorn, S.M. (1970). *J. Am. Ceram. Soc.*, **53**, 543.

Zener, C. (1934). A theory of the electrical breakdown of solid dielectrics, *Proc. R. Soc. London*, **145**, 523.

Zhurkov, S.N. and Tomashevskii, E.E. (1966). In *Physical Basis of Yield and Fracture*, Institute of Physics, Conf. Ser. 1, p. 200. Oxford: Institute of Physics.

Index